# PLAGIARISM: WHO REALLY CREATED WTC SKYSCRAPER DESIGN?

### By Greg Castle

NORDHAMMER PUBLISHERS

Copyright 2015, Greg Castle
All Rights Reserved by Author

ISBN 978-1-329-09996-8

# PLAGIARISM: WHO REALLY CREATED THE WTC SKYSCRAPER DESIGN?

## By Greg Castle

| | |
|---|---|
| STOLEN WTC SKYSCRAPER DESIGN | PAGE 1 |
| TO WHOM IT MAY CONCERN | PAGE 32 |
| SACRED GEOMETRY | PAGE 36 |
| GENERAL THEORY OF FUSION PHYSICS | PAGE 45 |
| PENITENTIARY VS COMMUNITY HABITATION | PAGE 52 |

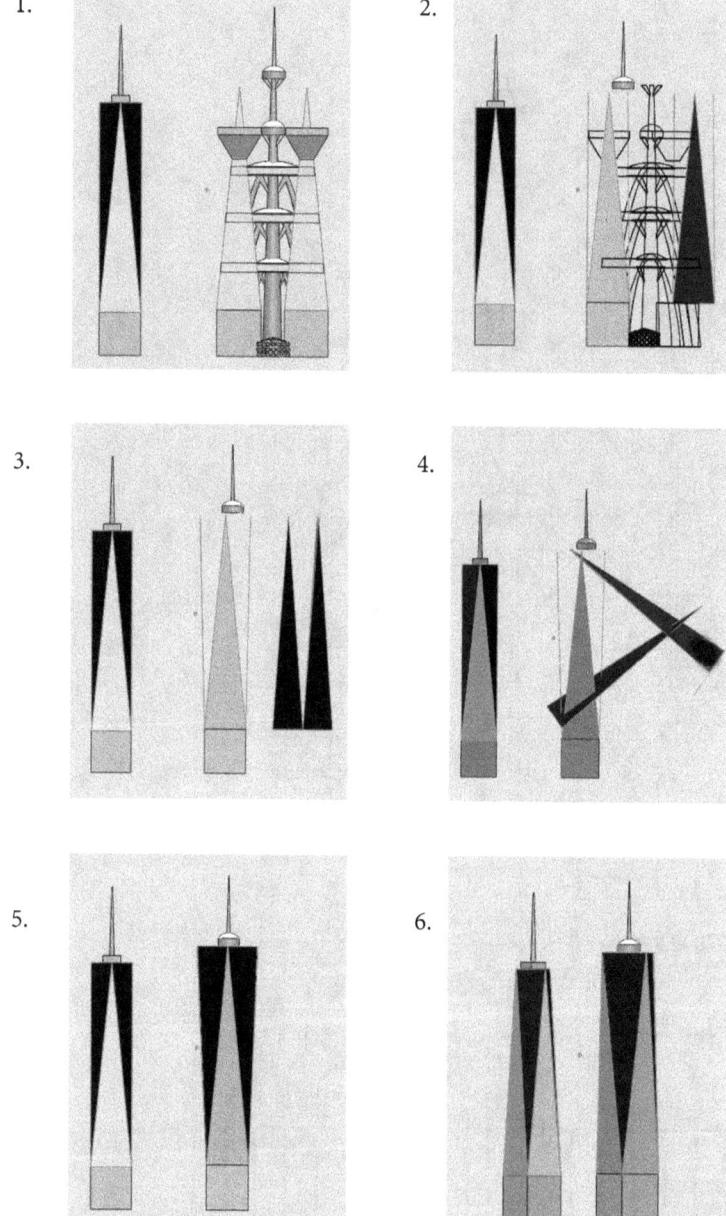

NEW WORLD TRADE CENTER: "TWIN OBELISK", COPYRIGHT, 2002 - (AUTHOR)ARCHITECTURAL, "COMPARATIVE ANALYSIS", AGAINST COMPLETED, WTC DESIGN: "COMPOUND OBELISK": SUGGESTING, THE MODUS OPERANDI, OF "NUMEROUS STRIKING DESIGN/STRUCTURAL, SIMILARITIES/ELEMENTS" TO ORIGINAL 2002, "WTC TWIN, SKYSCRAPER CONCEPT", S.O.M., C.S.U.K., 2014,

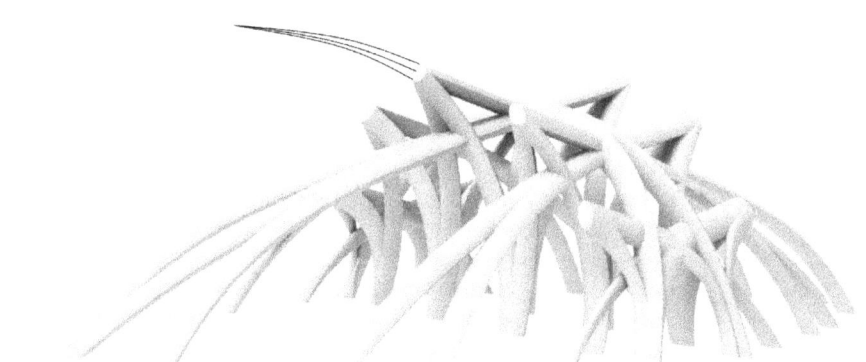

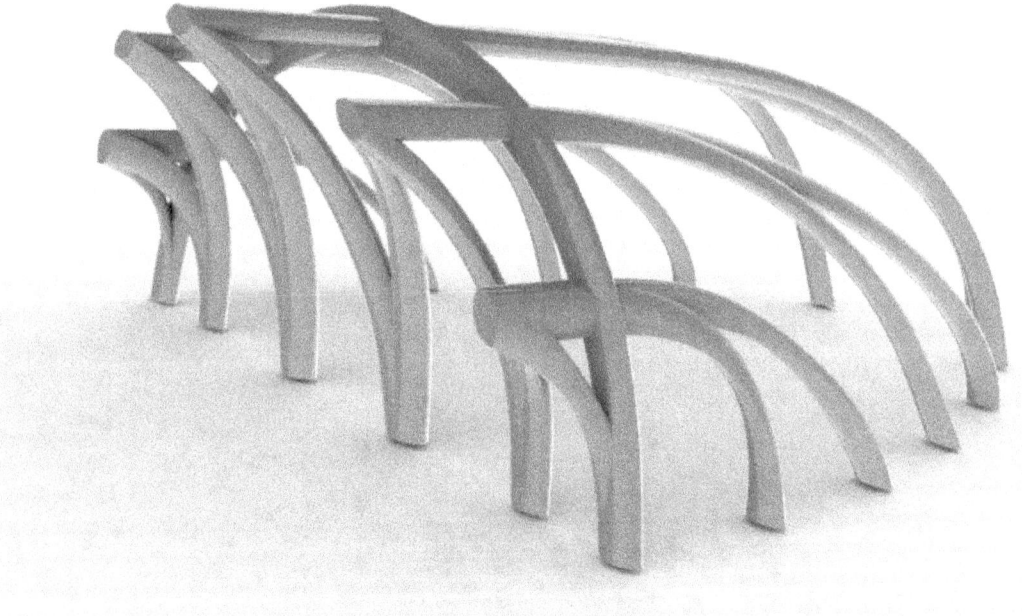

# Ground Zero TV tower OK with victim-kin activist

By LOIS WEISS

The leader of a Sept. 11 victims group says she's willing to consider including a telecommunications observation tower as part of the memorial for the former World Trade Center site.

Monica Iken, whose husband perished on the 84th floor of the south tower, told The Post she would not object to a tower as tall as the former WTC buildings.

"If they build a tower to show the height, and it was a communications tower and had an observation deck, that would be fine," she said.

"As long as it was unoccupied and not an office building," Iken added.

The mast destroyed on top of the north tower stood 360 feet tall and supported 10 main television antennas and a host of auxiliary antennas.

A consortium of telecommunications executives is trying to find suitable place to build a replacement while staying out of the path of airplanes.

Mayor Bloomberg has quashed the idea of putting a new tower on Governors Island, while other New York politicians are worried about losing the antenna to New Jersey, where officials are considering a scheme for a broadcast tower on Liberty Science Museum grounds.

As reported in The Post last month, one proposal for Ground Zero includes an up-to-2,000-foot observation tower and restaurant.

# STOLEN WTC SKYSCRAPER DESIGN:

Conspicuous Silence, Upon a Colossal Matter, Speaks Volumes, to a Collusion of Conspiracy - The Philosophical Discourse of Pure Reason, By Intrinsic Nature, Will Not Allow To Stand, Without Thorough Dissertation, Researched Analysis, Vetted Hypothesis, That Corresponds Directly, to the Reality of Things - To Sit Idly By, and Bare Silent Witness, to a Lapse of Fact, is an Untenable Act, When One Has Critical Means, To Disclose and Define, The Vain Impetuosity and Intellectual Sloth, Hegemony of Big Lie, In Effect, By Contributing Nothing, In Way of Dissenting Voice: Become Part and Parcel, Of the Original, Tissue of Lies, Contributing to the Propagandist Fallacy: We Would Be Made To Believe, as The Facile Truth - This Tenor of the Age, This Impulse to Glorify, This Obnoxious Prevailing Impulse, to Fabricate Reality, is Repugnant, in it's Inherent Nature, And Lessens Us, in Our Cultural Vitality, Diminishes Our Moral Understanding, and Invites a Plethora, of Larger Grandiose Dissimulations, Perpetrated Constantly Upon Us - Undermining the Foundation and Keystone, of Our Communal Polity, Deteriorating Trust and Fraternity - If One Has Means, To Speak Upon Such Things, With Authority, Upon Such Matters of Great Import, They Must Be Written Bold And Large, Upon Such Spurious Subjects, When At Hand - Compelled To Do So, When Times of Corruption And Decadence, Would Hold Truth, At Bay, With Damoclean Sword, In Annihilating Sway - Upon This Central Premise, I Will State My Considerable Case, A Perfect Didactic Example, That I Will Bring To Light, Within This Volume - To Bulletize And Recordate: What The Powers That Be, Would Seek To Conceal, Propagandize, Collectively Conspire To Compromise, To Maliciously Obfuscate, The Facts Compiled Here, For Anyones Circumspect Perusal -

As an Author and Inventor, of the Matter At Hand - I Have Unusual Perspective Upon It, Would, Set It Down Thus, Freedom of Speech, is Fundamental, When a Plagiarism Of Intellectual Work, Has Been Blatantly Committed - The Facts, As They Will Be Presented, Are Self Explanatory, You Become The Judge and Jury, To It All - Research, Will Sound Them Out, as Things of Verity - Silent Upon It No Longer, Bringing Forth The Compelling Evidence, of Stated Reality - I Would Have The Truth Be Told, Nothing More Is My Ambition Here - No Recourse Left, at My Disposal, Than Making Matter of Fact, My Philosophical Citadel - I Invite Those, Who Have Trespassed Against Me, To Make Counter Claim, Better Yet, To Thus Serve My Purpose - I Welcome The Fiction, They Would Fabricate - I Suspect, They Will Remain Conspicuously Silent, On The Matter At Hand, Retreat Into Professional Reclusion, When Posed, the Investigative Questions As The Evidence, of The Collective Plagiarism, Becomes a Matter of Open Discussion, Let Them Cringe and Tremble - Let History and Time, Correct This Factual Wrong, Truth is On My Side, The Testimony, Is Brought into Light, The Evidence Will Be Made Clear, The Bold Facts, Sluiced From The Prevailing Dross - Let Them Say What They May, I Will Not Be Intimidated, or Manipulated, Into Silenced Fear, Too Long They Have Tried This Before, To No Avail - It Is Now My Turn, To Speak Volumes Upon It - Let Their Egotistical Lies, Be Damned - The Emperor Has No Clothes, Let Us Mock Him Naked

This Philosophical Tractate, Also Meant To Edify, Create Provenance and Reason, For The Architectural Invention, I Have Wrought - A Thing, is Not Created, in an Intellectual Void, Conceptual Vacuum, It Requires Systematic Context, Engineering Genius, Aesthetic Discernment, Polymath Knowledge, To Be Brought Fully Conceptualized, Into Being - Few Have Had, Such a Rare Opportunity, To Philosophize Thus, It Requires An Encyclopedic Mind, With Many Skill Sets, To Formate Such a Perfect Storm - Fortunately Graced, With Multiple Abilities, I Can At Once, Design, Build, Create, and Write Eloquently, Upon The Philosophical Matter, At Hand: For When is Plagiarism Ever Justified, Intellectual Work, Blatantly Trespassed Upon - What Does It Say, Upon The State Of Society, in Moral and Ethical Decline, When Such Intellectual Crimes, are Commonly Perpetrated, Without Remorse or Seeming Consequence - And of the Original Artist, the Creator of a Thing, Must He Go Uncompensated, Unrewarded, for His Epic Endeavor, Unjustly Highjacked, From The Fruits of His Labor - Insult to Injury, At The Caprice, of Devouring Oligarchies - Where The Self Justifying Mantra, of Purest Greed, Defines The Age, We Live In - Concentration of Wealth, In All Forms Stolen, Hoarded In Dragons Lair: Collective Larceny And Technological Theft, of Intellectual Innovation, Determines Everything - The Warbled Tenor, Beggars Opera, of These Troubled Times, Economic Chauvinism, That Might Makes Right, Antithesis to Renaissance of Free Thinking - The Professional Role, of the Creative Artist, Eroded Into Anonymous Servitude - The Impeded Flow of Innovation, Silenced Philosophical Discourse, Mined For Propagandist Use, Culled of Data, Made Void of Ethical Reason With Programmed Dumbification: Extinction of Progressive Knowledge, Made Vestigial

Having Experienced Such Things, at Closest Range, Labyrinthian Mazing Corners, Victimized by Such Onerous Mentality, Then Cast Off, Like Debris - Intellectual Genius Commodified, Objectified and Quantified, Taken Beyond, the Original Individuating Genius - Most Dangerous Precedent, Ludicrous, to Culturally Embrace, Encouraging Brain Drain, and Intellectual Depletion, Through Mass Exodus Emigration, To Greener Pastures - The Robber Barons, Thus Robbing Themselves, of Their Greatest Resource, Eventually Running Out of New Ideas, to Steal - Disincentive to Invent and Innovate, Takes Hold - As News Spreads, Of Such Epidemic Infringement, News Travels Far And Wide - Symptomatic of a New Dark Age, Looming Upon the Horizon - Intellectually Cannibalizing Upon Each Other, Rather Than Individually Create, Genius Impulse Gone

The Perpetrators, Know Well Their Crime, Have Thus Named Them All, For The Public And Historical Record, Preserved - Their Complicity and Common Knowledge, of This Plagiarism, Will Not Escape Detection, in Greatest Annals, of Intellectual Theft, Their Names Writ Large, Forevermore - Let Them Bathe and Bask, In Their Newfound Notoriety - Shedding New Light, Upon What, Is Not Theirs, To Rightfully Claim - Let Them Explain, As I Will Do, The Genius and The Provenance, The Innovation and Architectural Engineering, That Went Into, This Revolutionary Work - In Fact, I Dare Them, To Explain Themselves, in Public Forum - To State Their Fallacious Case, With Forked Tongue Wagging - As Fumbling Actor, With Unlearned Lines, Addresses an Audience, Expecting Shakespearean Monologue - The Effect More Comedy, Than Tragedy - Although an ACTUAL Tragedy, Has Been Committed, HAS Occurred, None

The Less - Having Recourse, to the Public Wheel, The Free Press, Alerting The Most
Discerning, Among Us, Who Would Know Such Things, As Fictitious, or Actual Reality
Generally Sickened, by the Arbitrary Complacency, of Such Culpable Crimes, That
Accelerate, our Moral Descent, Breach of Ethics, Into Moral Bankruptcy, Of Spiraling
Cultural Decline - Architecture, Not Intended As Ivory Tower, Narcissistic Bauble, For
Real Estate Cabal - Like Randian Philosophy, Novel Expressed Form: "The
Fountainhead", Delineates, Level of Cultivated Civilization, Degree of Aesthetic Science,
Progress in Engineering, A Thing Unto Itself, a Product of Contemplative Genius - No
Authored Confusion, There Allowed, With Those Who Would Counterfeit, Stolen Design
Amend, Or Steal It Wholesale, To Be Disgraced in Imitation, When Mysterious Truth, Be
Found: Mind Over Matter, Triumph Over Ignorance, Truth Is Indomitable Spirit, Unbound

Removing, The Obstructive Wooden Sabot(eur), From the Machinery of Progress,
Is Tantamount Pinnacle, To Human Evolution, Ending General Strike, of Pure Reason -
The Means of Production, Becomes Autodidactic, Through Experimental Creativity,
The Innate Necessity, of Increasing Scale of Efficiency, The Decrease of Diminishing
Return, The Mother Science, of Natural Selection, Invention's Paradigm, Genius of
Original Prototype, The Greatest Metaphor - Those Who Would Hinder, it's Perpetual
Machinery, The Brilliant Imagination, Direct Byproduct, of The Inquiring Mind -
Purloining Of Intellectual Property, Through Patent Infringement, and Flagrant
Plagiarism, are Anathema, And Abomination, To Such Progress - Their Stolen Wares
Imperfect, Unwholesome, Tattered, Misconceptions - No Provenance or Intellectual
Labor, Informs The Aborted Original Work, Raped, Picked Clean of Invention, Then
Discarded: Unwarranted Credit Bestowed, Misplaced Authorship Debased, Breaks The
Causal Steps, Towards Original, Direct Conception - Counterfeit Forgery, Termination of
Progressive Line, of Subtle Reasoning, Brings The Scientific Momentum, of Empirically
Extrapolated Philosophy, To Mindless Grinding Halt: Until Saboteurs, Caught and Found

Thus Amalgamating, The Original Sin, Of Intellectual Trespass, Destroying The Creative
Pathway, To Originality, With Phantom Approximations, of The Authentic and Novel
Work - Unexplainable, Pseudo-Creation, Void of Rigorous Mental Process, Critically
Immune, To Further Intellection, Research, Investigation - Then Replicated,
Ceaselessly, into Premature Obsolescence: Withering Like Tainted Fruit, Upon Rootless
Synthetic Vine, Thorn Overgrown, Turning Poisonous - Aesthetically Deemed,
Unpalatable, As Freakish Mutation, That Cannot Be Fathomed, By Intellectual Thought
Trajectory, Readily Defined, or Phylum Categorized: An Artificially Euthanized
Contrivance, a Lifeless Uncontemplated Thing, Plastic and Worthless - What
Satisfaction, Can Be Derived From This, To The Rank Plagiarist Architect of Record,
Or the Suspicious, Aesthetic Observer - Whose Rebuffed Inquiries, Into The Work's
Origin, Are Met, With Unprovable, Untenable Assertions, Egotistical Scorning Hubris:
Probing Questions, Left Unanswered, With Conspicuous Silence, Covering Their
Original Crime, With Subsequent Layers, of Babbling Dissimulation - Speaking With
Riddles Of the Sphinx, Lying Game of Proxy, Childish Renditions, of Truth or Gambled
Consequence: Sophist Language Parody, Explaining NOTHING Resembling Substance

Professional Reputations, in the Creative Arts, Hinge Upon Original Works Creativity, NOT Stolen, Butchered, Knockoff Imitation - Artifice and Fashion, Cannot Camouflage Deception - Factual Reality, Has Logical Sequence, Adheres to Principles of Reason, Leaves Connecting Trails of Evidence, Born Witness Too: Are Unique and Open, By Intrinsic Nature, Have No Need, To Obscure or Hide, Their Creative Origin or Process, Resonate With High Fidelity, Are Extensions, and Projections, of the Creative Mind, They Were Born From - Thus So Explained, In Their Purpose and Intent, By Their Original Creator, Easily - That Crystal Clear, Coherent Philosophy, Informs The Work, of Skyscraper Architecture - To Give Rhyme To Reason, Eliminating Skeptic Doubt, Specify, The Unique Methodology, Intended For Architecture, of Such Pernicious Height I Owe Myself, At Least The Final Satisfaction, That The Architectural System Advanced, Brought Into Question, Will Not Be Lost, in Plagiaristic Butchery - And Resounding Silence, of Copyist Ignorance - Provenance of Creative Record, Preserved And Fully Published Here, The Philosophy and Unadulterated Schematics, Untainted and Intact

Like Ancient Archeology, Buried Under Accumulating Sediment of Time - Truth Reveals Itself, As To The Origin of Things, Eventually The Facts Are Discovered, Through Excavation - I Too, Say This Boldly, And Thus Provide, The Overwhelming Evidence, Gathered in Arduous Designing Process, From Dream Conception, To Superturnkey Engineering - Skyscraper Architecture, of Such Scale and Magnitude, Must Be Constructed, Like a Complex Machine - All The Forces Of Nature Calculated, To Withstand Amplified Forces, At Such Height, Huge Weight and Volumetric Mass - Vertical Structure Designed, At Such Tremendous Height, Requires Interdisciplinary Sciences, Beyond The Realm of Architecture Alone, To Break The Previous, Vertical Constraints, Of Previous Skyscraper, Building Methodologies - This is the Sticking Point, That Cannot Be Made Counterfeit, No Lapse of Knowledge, Can Imitate This, Through Faulty Engineering, As A Thing, Unto Itself, It Stands Alone - Remove Any Design Component, The Architectural Skyscraper Machinery, Interdependent/Integrated Dynamic Nature, Is Physically Broken, Causing All The Elements, To Cease To Function

Faulty Redesign and Material Cheapening, Reintroduce, Vertigo Sidereal Sway, Weakens The Obelisk, To Concaving Forces, Endemic, in Unsupported Pyramidical Structures: Without Internal, Upthrusting, Counterforce Buttressing - Intersecting Structures, Designed to Address Such Gravitational Stressors, Within the Multifaceted Rubric, Of a Pythagorean/Euclidean, Multi-Dimensional Plane - To Maximize Sway Stability, and Functionality, of Interior Space, While Enhancing Aesthetic, Architectural Beauty, With Exterior Geometric Harmony, First Form Structural Integrity, Titanic Load Baring, Maximizing Synergistic Strength: The Ultimate Purpose, Herein Lies, Rendered Transparently Clear, Made Obvious - With Eloquent Articulation, of Design Program Philosophy, That Only The Real Inventor, of Realized Conception, Can Actually Explain: With the Unique Signature, of Empirically Gleaned Discovery, Self Evidenced Truth, Stated, in Provable Fact, No Confabulated Fiction, Can Render Justice - Read On, and You Shall Find, This Meaning Here: The Greatest Crime, of Intellectual Theft, Patent Infringement, And Architectural Plagiarism, Ever Committed - I Would Purge, This Intellectual Trespass, Philosophically, From My Creative Consciousness - Thus Leave

Behind, Cautionary Tale, of Abused Good Will, To ALL CREATIVE Individuals, Not Be Seduced, By Unchecked, Unregulated, Intellectual and Creative, "Corporate Interest" Piracy - It Is For You, I Intended To Write This, May It Guard You Well, From Repetition, Of Plagiarism, and Creative Piracy - Raise The General Awareness, Among Fellow Creative Artists, Pass The Message Forward, Protect Yourself, Spread The Portent Warning Words, Do Not Have Your Good Will, Violated: Remain True In Creative Vision

This Design Chronicle, Sets The Story Straight, Beyond Blighting Conspiracy, Establishing, The Original and Unique, Design Philosophy, That Attended, the Innovative, Architectural Skyscraper Conception - So That It Be, Philosophically Salvaged, Through Unexpurgated Dissertation, Informed by Knowledge, Verified in Provenance, The Truth, Of It's Mysterious Origin Solved, By Testimonial and Designers Memoir, Completely Explained, Enduring Official Record, of it's Architectural History, From Start to Finish - My Silence Upon This, Has Been Broken, In Contemplative Retrospection, Having Gathered My Thoughts, Upon This Complex Subject - Exercising My Fundamental Right, To Free Speech, In Light of, Oligarchic Prevailing Tyranny - My Voice Will Be Heard, Upon The Matter, My Conscience Freed, From Silent Angst Speculation, Rendered Thus, Upon The Page - To Lying Detractors, Egotistical Dilettantes, Sychophants, Architectural Impostors, Clownish, Stand Ins, The Truth Will Define, Your Hammish Thespian Character, In The End - Vainglorious Repugnant Acts, Stolen Thunder, Will Soon Haunt You, Your Actual Felonious Profession Revealed, To Be That, of Fraudulent Cheater, Pseudo-Intellectual Thief, Most Contemptuous Liar

Be That As It May, Your Fate Concerns Me Not - As Your Total Disregard, For Your Theft From Me - Fully Warrants, My Dedication, Of This Expose Book, To You - Feel Free to Attempt To Plagiarize and Further Warp It's Meaning, As Well - Habitual Offenders, All - Recidivist, Never an Original Thought, Invades Your Collective Ignorance, In Mundane Malingering Sloth - Match Wits With Me, If You Dare, Try As You May - I Am Formidable, And More Than Welcome, The Continued Challenge, To Expose Your Bold Impostory, To The Unsuspecting World - As Matter Of Stated Fact, It Would Amuse Me, Immensely

You Are Now, The Grist of My Poison Pen, No Skyscraper Design To Steal, This Time Around - Outside The Comfort Zone, of Your Preferred Form, of Grand Larceny - War Of Words, Is Now Proposed - Fitting Subject, For A Book, I Could Not Resist To Write, To Catalog Your Horrendous Crime - Raconteur, As I Am, With An Extraordinarily, Uncanny Memory, For Complete Detail, Wishing To Further Memorialize, Your Invasive Meddling, Into My Creative Life - And Derive New Wellspring of Creativity, Written Form Inspired, Ironically From Transgression - Thus Justified, To Make Righteous Mockery, Blundered Dubious, Purloined Design, Undeserved Notoriety and Fame - Owning The Skyscraper Design, You Have Stolen From Me - The Compiled Truth, of It's Architectural Inception, Explained - I Will Now Toy With You, To My Hearts Content, Like Cat With Fleeing Mouse, No Where, To Run And Hide: Made Provocateur, in The End, With Dissenting Voice, Echoing in The Wilderness, I Shall Call You Liar! - A Blatant Sham Perpetrated, Which Stands Apparent, Upon the Skyline, Aborted In Facile Form - I Set My Credulous Mark, Upon You Now, Fire Branding You, Malicious Thievery, With Scarified Letter,

Upon Your Thoughtless Brow - Wear It Well, With All The Indignity, It Then May Offer,
For All To See, And Rebuke You: Laughing At Transparent Crime, Telegraphing Boldly

Sardonically Made, Into Public Laughing Spectacle - Betraying Your Chosen Profession,
Simply Labeled, Thieving Plagiarist: Read On In Shame, With Lowered Scarified,
Forehead - Espy All Your Names included, See, If I Have Missed Anyone, or Anything,
At All - Thus I Vent My Spleen, Upon You Now, With Accumulated Rage, Turned
Venomous, With Long Silence Broken, Watch The Other Shoe Falling, Lawyer Up, And
Run For Cover - All You Have Ever Done Previously, Made Suspect Now - With Farcical
Assertions, Untarnished Reputation, Growing Despicable, In The Discerning Public Eye
Corrupting Forces, Create Reforming Movements - The Ignoble Die, With Their Names
Buried With Them - No Tears Are Shed, For Specious Liars: Even Their Paltry Eulogies,
Become Grandiose Fabrications, Tissue Of Invented Lies, From Pretentious Friendships
Interred, With Deadened Stolen Ideas, Contributing Nothing of Real Substance, With
Inconsequential Wake - I Offer This Up To You Instead, A More Accurate Assessment:
Multitude of Vainglorious Foibles, Taken To The Grave - Rarely Do I Indulge, in Vitriolic,
As My Body of Literature, Attests - However There Comes A Point, When Tyrannized,
One Must Take The Warriors Stance, NEVER Turning, The Other Cheek Instead - Such
Violation, of My Intellectual Rights, Has Come To Pass - Citizen, in a Nation of Laws, No
Obeisance Given, to Pretender Kings - Here Would Espouse, Such Revelations Freely,
The Final Bastion, of Philosophical Speech - Such Level of Indignation, Cannot Be
Tolerated, And Justifiably Call Oneself, A Man: Fighting Every Inch, Born Warrior Stock

Not Bound To Serve, Artificially Contrived Power - Born Free To Create and Say,
What I Please: Freedom is a Precious Right, That Shall Not Be Taken, Away From Me -
Inalienable, By Constitutional Statues, Insured By Civil Right, And Due Process of Law -
Plagiarism, Must Be Routed Out, In Free Republics, If The Great Experiment, Of
Democracy, Is To Endure - Political Concepts Formed, From The Brilliant, Untethered
Minds, of Free Men: Constituted To Assure Philosophical Discourse, Lawful Assembly,
Protection Of TRUTH: Invented, With Checks and Balances, To Stave Off Corrupting
Influence, Attempted Oligarchic Overthrow, Civil Insurrection, The Unjust Draconian
Hell, Of Arbitrary Unequal Laws - Some Would Have Us Retrograde, To Feudalistic
Pandering, Imprisoned Bondsmen, To Entitled Land, Imprisoned Feudal World - Servile
To The Master, of The Manor Born - Eviscerating, What Founding Fathers Proscribed,
When First Devising, Revolutionary Political System, To Liberate, From Aristocratic,
Colonized Subjugation - Many Among Them, Were Inventors Too, They Would
Sympathize, With My Copyright Infringement Plight, And Encourage Me, To Stand Up
For My Rights, of Free Citizenry, In Likewise Tradition, of The Original Patriots:
Continental Army Regulars, Militia, Rangers and Minute Men, I Encourage You To Fight

Read The Portentous Writing, Upon the Wall, My Story Speaks Volumes, To The
Growing Plague, of Elitist Tyranny - Stripped of Our Rights, Reduced to Chattel, Led
Down, the Primrose Path To Hell, Denuded of Our Intellectual and Material Valuables,
Unprepared, And Made Defenseless, For Holocaustic Sacrifice - The Nightmare Begins,
With The Lawless Anarchy, of Unchecked Pillaging - What Gives The Right, Self

Conferred, Upon The Powerful, to Steal What Is Not Theirs, With Abject Impunity - What Does It Say, Of Disintegrating Society - How Have We Arrived, At Such Vile Destination, Who Beest, The Pirate Captain, Who Pressed Us Into Service, Enslaved To Our Unknown Precarious Destination, Navigation Void of Course, To Slavers Port - I Would Know Such Things, And Speak Freely Upon Them, Share Clear Insight, Upon Dubious Matters Untold - Long Have I Tried, By Diplomatic Means, To Settle, This Copyright Infringement - With No End In Sight: I Have Taken It Now, To You The Conscious Reader, While Free Men Still Communicate: Writing With Renewed Philosophical Conviction, Forbearing Reluctance With Vigor, Casting Caution To The Wind, Let Freedom Ring, And Bold Truth, Be Daringly Told, The Silence Broken, Forevermore

Like Miltonic, Areopagitica, I Will Exercise My Civil Right, Publication Of Free Thought Elucidating a Matter of Fact, Revealing a Conundrum, Blatant Scandalous Lies Redemption Through Purgation, Casting Off Slave Yoke of Anonymity - Beyond the Overreaching Grasp, of Truth Obscuring, Purloining Cabal - Conspiracy of Silence, Would Have Hermetic Secrecy, Truth Censured Conclave, Prevailing Power as Law of The Land - Maintaining Insolent Lie, Against Generative, Root Cause, Creativity Denied, Original Meritocracy - Genius Accomplishments Concealed, The Stolen Works And Deeds Of Other, Made Vain Trophies of Looted Piracy - Vainglorious Narcissistic Impulse, To Dominate an Enslave, What is Not Theirs To Control - Intellectual Works, Left Uncompensated: Usury of Such Kind Unfathomable, in More Enlightened Times - Where Innovation and Invention, Was Attributed and Relegated, to the Sacrosanct Province, Of Individual Conceptualist Mind - Such Human Stepping Stones, Civil Right Stricken From Them - Become Labyrinthine Pathway, Minotaur Mazing, Sacrificial Hell - I State This Plainly, Not Just For Myself - But For Those Unsung Creators, Throughout History: Where Good Will Abused, By The Uncreative, Nay, Destructive Tyrant - Will Seek Out The Genius, Like a Form of Wild Game, Intellectually Poaching - Loathing Contemptuous, of The Original Creator of the Work, Yet Obsessed, With the Inspired Miracle, of The Creative Process, Denied Them - Many of These Creative Souls, Have Been Brought, to Conspired Ruination - Once the Tenderest Victuals, of Laborious Enterprise, Have Been Partially, Or Wholly Glut Consumed - Their Names Numerous, Now Nearly Forgotten, Their Pirated Discoveries, Often Laid To Waste, Made Faulty, In The Dubious Counterfeit - Alacrity of Theft, Obscures Purest Vision, From Truest Intent -

Thus Holding Back, Collective Progress, of the Human Race, By Petty Controlling Few, Inventions Lost, That Would Advance Us - Such Divine Impulse To Create, Often Met, With Murdering, of The Graced Messenger, After Ensign Deliverance, of Highest Truth - Extinguishing The Promethean Fire, Seemingly Pandemic, in These Turbulent Times - Where Concentration of Wealth and Power - Reliant on the Suppression and Torturous Exploitation, Of Creative Individual - Institutionalized Social Mortification, Proves Tried And True Methodology, Building Imperialistic, Corporate Empires, Upon Cultural Desert, Of Shifting Sand - Requiring Suppression and Theft Exploited, Bungled New Technologies, From Where And Whence, They Come And Go, Entirely Dominated, By Unchecked Greed - I Know Well, From Where I Speak - It Has Happened To Me Many Times Before - Yet Will Confine My Argument, To Skyscraper Architecture Only -

My Other List Of Inventions, Being of a Highly Secret Nature, Not Meant For Public
Consumption, Nor Security Breached, Further Discussion, Beyond Civilian Application,
Deemed Irrelevant, In Context of Civilian Skyscraper Design, Thus I Confine My
Argument - Enough Has Been Said Upon It - Suffice It To Say, I Have Been Robbed -

In My Own Particular Circumstance, Tapped For a Particular Skill Set of Abilities
To Develop and Design, The Largest and Most Stable Skyscraper, Ever Built - Known
For Superturnkey Engineering, With Patentable Applications, in Constructing, a Trolley
Line Prototype, Eliminating the Need, for Segregated Power Stations, Hard Wired Into
The Existing Power System, Therefore Having No Environmental Impact, Upon City
Infrastructure, Including the Elimination, of Earth Ground Electrolysis, Of Water And
Sewer Pipes, No Corrosion, of Underground Utility Systems (Common Destructive
Phenomena, of Early Street Cars and Trams) The Political Skills Required, To Navigate
And Maneuver, Complex Code Requirements, Of NYC Urban Design Review Boards -
Rightly Reckoning, My Unique Capacity, To Do So - A Skyscraper, Is A Vertical Form, of
Railroad Infrastructure, Which I Was Qualified, To Design/Build, in New York State:
Like Cooper Union School, of Fine Art and Architecture - Frame Built With Carnegie
Train Rail: Internal Beam Structure, Affixed To Stone Facade, The First Steel
Skyscraper Born, From Collaborating, RR Industrial"Robber Barons" - The Tallest
Building, Constructed, at that time, In New York City, Never Deviating, From That
Construction Paradigm, Disregarding Ergonomic Scale, Oblivious Of Altitudinal Forces:
When Peter Cooper's, Tom Thumb Steam Locomotive, Was Cutting Edge Technology

That Is, Until I Invented, An Isotonically Stressed, Unswaying, Dynamically Sprung,
Glass and Steel Variant - Specifically Designed, To Arrest and Neutralize, Gravitational
General Dynamics, in a Skyscraper Design Modulus, Meant To Rise Above, 2,000 Feet
Having Studied, Pythagorean, Euclidean, Cartesian, Newtonian Geometry - Extensive
Knowledge of Architecture, and Art History - Then, Able To Synthesize, an "Engineered
Machine Building" - Abandoning, the Obsolete Architecture, of the Past Century,
Post Lintel Absurdities, Windage Flexing Variations - Unable to Withstand, Geological
Physics, Imposed Upon Them - At Such Towering Magnitudes of Height, Occupying
Unique Environmental Space, Prevalent Natural Forces Existing, At Half Mile, Of
Atmospheric Altitude - I Went About My Business, Creating Architecturally Dynamic,
Structural Counterforce, Eliminate Sidereal Skyscraper Sway, Horizontal Kinetic Motion,
With Levered Compression, Euclidean Axis Point, Torc Geometry, Radiating from the
"Deadman" Tie Back, Building Foundation, Along Transept Beam, Cubical Points,
Tethered, to Centralized Reinforced Steel Concrete, "Bunker" Elevator Bank Shaft -
The Coefficient, Buttressed Counterthrust, Sufficient to Eliminate, Practically All Sway
Motion, Kinetic Sidereal Action, Throughout The Entire Structure: Was Subsequently
Verified, in The Hurricane Wind Testing Simulator, Architectural Scale Model,
Skyscraper Facility, At the University of Chicago - Test Records Permanently Archived
My Original Design Eliminated, Typical Twelve Foot Sway, Endemic Characteristic,
In Previous Skyscraper Designs, of Similar Mass and Similar Height - The Greatest
Offender, Previous Twin Trade Center, Pre-911, Designed With, Gravity Fed, Five Story
Water Tower, Which Caused Accentuated, Sidereal Sway Dynamics, Beyond the Twelve

Foot, Legal Limit - Subsequently, Such Water Tower Construction, Was Banned, in Skyscraper Design, As Structural Hazard - Additionally, The External Exoskeleton, Y-Shaped, Steel Beam Array, Meant To Be De-Constructed, Dismantled and Sent Down, Central Elevator Shaft, of The Building, During Future, Anticipated Staged Demolition, Proved Structural Weak Point, When The Building, Was Targeted And Razed, by Terrorists, Capitalizing Upon The External Design Flaw, Exploiting, The Original "Disposability" of the Building, Designed Into It, At Architectural Conception - Knowing These Historical Facts, Inverting the Design Paradigm, Utilizing A, "Central Euclidian Helix", of Internal, and External Beam Work, Impervious to Collapse, From a Single Impact Strike Point - Which Had Previously Collapsed, the Weakly Interlocked Beam System, and Pendant Five Story Water Tower, Like A House Of Cards - The Pinch Point Of Commercial Jetliner Attack, Calculated To "Chop" The External Beam Work, Like an Axe Felling a Tree, Causing a Sequence Of Cascading Collapses, Along The Sidereal Line, During High Velocity, Inertial Impact Point - Superheating Of Construction Materials, Through Leaking Jet Fuel Combustion, Further Weakened - The Exoskeleton Steel Structures, Ability To Sustain, It's Own Weight, Soon Melting Point Compromised: The Laws of Newtonian Gravity, Accomplished The Rest, Gravitationally Accelerating Demolition, Destroying The Old Twin Trade Center -

Using an Aircraft, as a Missile, Was Not A New Idea - The Empire State Building, Sustained limited damage, when a Commercial Airliner, Accidentally crashed into it - Due to the Steel and Rivet, Cubed Interior Construction, With a Limestone Exterior Edifice, it was able to withstand, The crash impact, without Destroying the Entire Skyscraper - So This Example of Structural Durability, Seemed The Most Logical Point, For Design Modality Departure: Analysis And Development, of Structurally Hardened, Terrorist Proof, Architectural Systems, That Could Be Built, at the 2000/2400 Foot Level, Of Vertical Altitude - Obviously the Ultimate Goal, Was To Redesign, The Original Set of Flaws, Out of the Original Skyscraper Architecture - So That a National Tragedy, Such As 911, Would Never Happen Again - It Was In This Spirit, My Expertise Was Solicited, Little Did I Know, My Good Will, Was To Be Totally Exploited, My Income Producing Business Ruined, In a Series of Occurrences, That I Will Clearing Indicate, and Piece Together, into RICO Arc - The cc List Provided, in the BHRA/DIG Letter Below, is an Open Rogues Directory, for the Truth Seeking and Inquisitive, as To Where and How, To Research the Perpetrators, This Book Will Guide You To Them, The Originally Intention

Where, When and By Whom, The Architectural Plagiarism, Was Originally Perpetrated - Providing Supporting Documents, Schematics, Diagrams, Copyright Forms, Time Stamped Receipts, Chain Of Custody, For The Original WTC Designs, in Question - Establishing Authorship Claim, and Historical Record, of the Skyscraper Architecture, Defining Design Philosophy, That I Alone Invented - The Sheer Fact, That I Can Write a Book, Upon the Subject At Hand - Indicates Clearly, That I Am No Fraud, Impostor, Plagiarist, Confabulator, of Ridiculous Cover Stories - Let These Men Write Their Own Books, Describe The Mystical Method, of How They Came Upon My Novel Design - What Crystal Ball Was Then Employed - Where Are Other Examples, of A Logical Design Progression, in Their Stolen Architectural Portfolios - It Would Seem To Me,

They Would Speak Fluently Upon It, Thus Far, I Only Hear Plain, Crude Statement, Tenuous Authorship, Void of Detailed Description: Like The Cat, Who Swallowed, the 2000 Foot Canary - Looking Forward, To Their Rebuttal, To This Tome, Willing To Debate The Subtleties, of Architectural Design Authorship, Anyplace at Anytime -

Particularly, After Series, of Remarkable Synchronicity, Having Encountered Others, Who Have Witnessed and Confirmed, The Forensic Flow Chart, of Architectural Plagiarism, AUTHORSHIP THEFT - Glaring Disgrace, to Their Noble Profession, At Very Least - Criminal Infringement, of Intellectual Property Rights, Categorically Provable Certainty - Read On, And You Shall See It More Clearly - Not The First Time, Accusations, of Architectural Plagiarism, Were Filed Against, SOM, David Childs

I Grow Fatigued, at the Prattling of Such Idiots, and Thus Invite, An Open Public Forum, To Conclude, The Evidenced Truth, of the Matter - Fantastic Allegations, are NOT My Intention, Rather The Reality, of a Fantastic Crime, That Has Gone Unnoticed, Committed in Fraudulence - Worthy, of Thorough Circumspection, and Historical Correction - I Do Not Write a Novel Here - The Truth of the Magnitude, of Such a Thing Cannot Go Entirely Unnoticed, Too Many Threads Of Evidence, Weave a Tapestry of Lies - If I Have Misrepresented Myself, They Have Recourse, To The Law, So Be It - I Vigorously Exercise, My Right To Articulated Speech, In The Sacrosanct Medium, of Free Publication - I Will NOT Expurgate, Nor Retract, Single Letter From This Stated FACT - The Time For Inquisition, and Torture Compelled Confession, On The Dungeon Rack, Are Centuries Over Now, May They Never Return Again, In Drips and Drabs, As Our Society Condones, The Multifarious Felonies, of Corporatist Oligarchs: To Accumulate, Like Epidemiology, of Bubonic Plague, Upon Dispossessed Masses -

That Being Said, Carrying On, Undaunted, in My Intended Task - The Barbarity of Stolen Thoughts, Uncredited Invention, Servitude To Invisible Cabals, of Corrupting Powers, And Secret Influence - Is Fundamentally Odorous, In Free Society - We The People, Must Assert, Our Inalienable Collective Rights, Guaranteed, By The Framers Of The Constitution - Democratic Republic, Built on Laws, Not Piracy, Rape and Pillaging, Of Robber Baron Tyrants - Where Does The Lies Stop, and The Truth Begin, Are We So Far Beyond Reform, That Such Travesty of Justice, Sold To Highest Bidder, Becomes Malignant, Commonplace Mockery, To Pure Reason - This Is Not The Same Country, I Was Born, and Raised In, This Is Not The Land of Liberty, My Grandparents, Sought Out, To Find New World Sanctuary, From Culturally Stagnated, Wartorn Europe -

I Say This, With Heavy Heart, Victimized, By Such Crime As This - May It Serve As Example, Of Where We Are Heading, Both Legally and Politically - Destruction of Individual Civil Rights, In This Case, Architectural Plagiarism - Having Attempted Every Other Recourse, To No Avail - So I Leave It, To This Tractate, To Document Fully, And Memorialize My Tale - Let No Man, Stand Down, To Such Catalogs of Abuse, May This Book, Embolden You, To Seek Justice Out, Renumeration, Due Credit, Documented Authorship - The Creative Artist, Must Be Protected, From Such Unprofessional, Condescending Slights, Intellectual Purging, and Invention Thievery, Are the Signature

Hallmarks, of Totalitarian States - Is This The World, You would Have Your Children Inherit, Lambs To The Slaughter House, Driven Herd, of Anonymous Chattel - Let Us Not Slip This Occasion, To See Justice Rendered, Without Pandering, For Servile Mercy, To Restore What Is Ours Already - Through, The Third Estate, of Free Speech, Let Truth Be Told - Greatest Sanctuary, Of The Rights of Man, Take Solace and Refuge In It - Espousing Fact and Truth - Let The Proof Displayed, Disprove Conjecture, Deflate The Megalomaniacal Ego's, That Have Perpetrated, Intellectual Theft, Of The Highest Order: Their Criminal Story Be Told, Let History Be Judge, Favors, Intrepid and the Bold

I Will Not Stand, Any Longer, This Perpetration, of Creative Farce - I Am NOT A Charity Of Ideas, For Malingering SLOTH - Let My Accusations, Be Sounded Out - Enough Speculation Abounds, As To, Who Actually Designed, the WTC Skyscraper Architecture, The Transportation Hub, "The Scethlia Shard Building", London, UK - The Coney Island Redevelopment, EDC Master Plan - Would Go To Those, "Impostor, "Stand In's" To Put The Question, Plainly To Them - Look Them DEAD In The Eye, And See Them Cringe And Squirm Nervously, Attempting To Lie Badly, Right In My Face - Bring Them a Copy, Of This Book, Primer on the Architectural System, Originally Invented - So That They May Attempt To Plagiarize Again, in The Public Record, of Copyrighted Material - I Would Relish That - Quiz Them At Length, If They Have Any Concept, or Indolent Understanding, Of It - Beyond The Actual Theft of Plans, Remaining Mute And Clueless

Waiting, For The Other Shoe To Drop, Attempting Vainly To Guard, Their Imperiled, Professional Reputations - Of Which, I Do Not Care, They Are Disgusting, Wonton Parasites, To Me: Trespassers Upon My Intellectual Property - I Treat Them, With The Same Blatant Disregard, Reciprocal Disrespect - They Lavishly Expended Upon Me, To Bury My Authored Creation, Claiming It As Their Own, With Vain Sophistry: Scandal Mongering, Concept Hoarders: The Truth Has Come To Roost, I FULLY Intend For You, To Thoroughly Enjoy Yourself - Your Example, of Copyright Piracy, Has Inspired Me To Create Again - I Dedicate This Book To You - Without Your Criminality, I Would Have No Subject Matter, To Thus Chew Upon - I Will Restrict My Condemnation, To The Jury Of The Public Weal, Let Evidence Become Discovered Fact, Determine Innocence, or Obstructive Guilt - My Conscience Is Clear, My Words Ring True - Accomplishing This Effort, Satisfies My Psychological Need, To Rectify Most Vile Wrongdoing - Given Further Opportunity, Positive Sense of Accomplishment, Establishing Provenance, Design Philosophy, Original Architectural Aesthetics, Where a Plagiaristic Void, Has Been Created - To Further Show, the Actual Conception, of the Skyscraper Engineering, Denuded, of Camouflaging Artifice, Unexpurgated in Detail, Complete Design Program Concept Provenance Preserved, Engineered Invention, Paradigm Course Corrected

Vindication Within This Thesis, is Self Affirming, Divulged Imprinted Reality - Too Long, Remaining Silent, on the Matter - Frustrated, at Every Turn, To Right The Wrong - Conflagration Of Professional Betrayals, Conflicts of Interest, Worthy of Disbarment, Architectural Con-Artistry - Having Exhausted Myself, My Good Will Exploited, Moral Resolve Depleted, Soldiering Onward, As Critical Writer Now - I Owe It To Myself, To Compose an Accurate Accounting, Clear Historical Record, This Quantum Leap, in

Architectural Design, Skyscraper Engineering - Left Thus Unexplained, In The Current Butchered State, Polyglot of Haphazard Juxtapositions, Swaying In The Breeze, Like Autumnal Leaves - I Am Fully Within My Rights, To Aesthetically Criticize, What Could Have Been, The Greatest Manmade Structure, In The World: Beyond Insult To Injury, Final Analysis, Ruined Stolen Concept, Like Casting Pearls Before Swine - I Would

String Them Back, Again, Into Priceless Necklace Ornament - Methodically Documented Here, For Further Innovations, To Grow and Evolve From - Like An Aborted Child, I See The Plagiarized Remnants, Crying Out To Me, To Correct The Botched and Butchered, Architectural Design, Tampered Offspring, In Stark Comparison You Will See Them Too, And How The "Shell Game" of Altered Design, Was Cleverly Perpetrated - The Poverty Of Weak Ideas, Thus Recreating, The Fatal Design Flaw Characteristics, of The Original WTC Skyscraper Architecture, Reintroducing Destabilized, Accentuated Sidereal, Kinetic Sway, Condemned Midway, Through Constructed, For Having OVER Fourteen Foot Sway, Reinforced With Steel Gusseting

May This Thesis, Serve As Dunce Cap, for Charismatically Bypassed, Architects of NO ACCOUNT, Who Saw Fit, To Pilfer and Cherry Pick, a Design That Was Never Theirs - If These Epithets And Fitting Nomenclatures, Offend Their Vain Sensibilities, Better Still Look What You Have Stolen, Then Bungled, Hang Your Thoughtless Heads, in Public Shame: Lowest of The Low, Scum of the Earth, No Better Than Common Thieves, With License To Steal - I Have Nothing, To Loose BY THIS, and Everything TO GAIN, By The Revelation, Your Dissimulation Has Given Me, Windows of Opportunity, For Further Design Projects, And For This, I Thank You: Tables Are Turned, The Cast of Doubt Upon You, Explain Yourself, Or Silently Endure, My Verbal Thrashing - Amusing Me, To Flog You, In Your Transparent Guilt - Like Confederacy Of Dunces, You Have Outdone Yourselves - Evidence of Your Crime, of Architectural Plagiarism, 1776 Feet Tall - My Vitriolic Condemnations, Flow Like Poisoned Wine, Your Position Indefensible, The Pleasure Mine - Such Satirical Tractates, Where Often Written, In The Age of Enlightenment, To Persuade, Edify and Heighten Awareness, Through Targeting The Absurd, Ludicrous, Often Ridiculous, Outworn Thinking, Anachronistic Custom, of The Day - In The Critical Spirit, Of Voltaire, Montaigne, Emerson, Jefferson, Hamilton, Adapting Their Method of Philosophical Discourse, To Suit My Present Need, To Redress Such Crime - To Place Microscopic Lens, Upon You, For All To See - Like A Skyscraper Design, I Will Build My Argument Around You: I Guarantee, It Will Not Sway, But Certainly Persuade You, From Error Of Your Ways, Gained Satisfaction in Knowing, That I Am Grieving You, Revealing Who You REALLY ARE: Thieving, Despicable, Lying, Ravenous Wolf Pack, of Architectural Engineering, Turned Plagiarists - This is the Unfortunate, WHOLE Truth Upon The Matter, And NOTHING More -

Like Newtonian Calculus and Trigonometry, Calculating Graduated Acceleration Tables, Reaching Terminal Rates, Potentiated Descending Velocity, in Friction and Gravitation Based Cascading Atmospheres - Equilateral Fall, Of All Objects, Within Quantified Vertical Space - The Opposite and Inverse Set, of Physical Laws Apply, Within Vertical Skyscraper, Building Structures, Construction Modulus Formulations: Counteracting

Systems, Must Be Devised, To Neutralize, Arrest and Negate, Geometrically Downward, Increasing Structural Loadforce - Compensating Countermeasures, for Sidereal Windage, Across Massive, Aerodynamic Facia, Architectural Multi-Plane - Becoming Isotonically Static, Within Optimized Tolerances, of Intersecting Material Possibility, No Longer Vulnerable, To Typical Skyscraper Vertical And Sidereal Sway, Atmospheric And Gravitational Forces, Synergistically Generated - Obelisk Based Ratios, Reinforced

With Graduated Ziggurat, Stepped Pyramid, Interior Counter-Convection Buttressing, Creates Modulus of Structure, That Can Withstand Environmental Stressors, In Excess Of One Linear Mile - Twin Obelisk Structures, Isotonically Tethered, in Euclidean Sections, Can Be Scaled To Monumental Heights, When the Architectural Engineering, Becomes A Dynamic, Spring Loaded Mechanism, Counterbalancing Itself In Space, Creating Unswaying, Tensioned Homeostasis, Within Itself - Thermodynamic Friction Tolerances, Physical Tensile Energy Discharged - Is More Akin To Anatomical Skeletal Structure, Biomorphic Engine Paradigm - With Dynamic Self Adjusting, Stabilized Design, Beyond More Simplistic Structures, Design Obsoleted, in Recent Past -

Turning Instead To The Strongest and Most Enduring Architectural Structures, That Have Remained, Archeologically In Tact, For Aeons of Time, Synthesized, in an Elegant Intersecting, Engineered Design Mechanism, To Combine Strength, Lightness, Height, and Interior Volumetrics, in a Skyscraper, Without Kinetic Sway - The Terminus Height Implied, Microenvironment, of Potential Hurricane Wind Gust, Atmospheric Acceleration, Radiant Thermodynamic Heat, Thinning Altitudes of Atmosphere - Causing Material Convection Absorbing, Warping Materials, How To Counterbalance, These Instability Variables, Became Apparent, Once the Strongest Historical Architectural Systems Devised, Were Employed, in Synergistic Dynamic Intersections, Engineering Paradigm

The Egyptian Obelisk, the Babylonian Ziggurat, The Bridge Buttressing, Rainbow Truss Arch, The Geodesic Dome - All Known To Load Bare, Geometrical Magnitudes, Beyond Their Own, Size and Weight - Naturally Emulating, the Insect Ant, Also Equipped With Exoskeleton, Able To Carry Twelve Times, It's Own Mass, Relative To Size, Transporting Proportionately Huge, Load Baring Weight - Truly, a Wonder of Engineering, to Behold - Furthermore, Threefold Suspension Skybridges, Mounted Upon Three Matched Sets, Of Graduated Spanning Arches, Deadmanned, to the Skyscraper Atrium Foundation - The Sky Bridges Offering, Pre-Sprung Downward Compression, at Apogee, of the Arching Span Load - Furthered Stabilized, By Central Sky Needle Shaftway, Establishing Third Vertical Tethering, Axis Element, to the Overall Design Paradigm - Crowned by Two Geodesic Domes, One, a Spirally Restaurant - The Other, a 911 Memorial, Observation Tower, Above It - To Be Set, at 2400 Feet of Altitude - The Skyneedle, also serving, as An Additional External Elevator Bank, From Ground Level - The Glass and Steel Bridges, Offering Panoramic Skyline Views, Glass Encased Strip Mall, Boutique Commercial Spaces - Self Contained Urban Development, All the Necessities, of Planned Community, Fully Provided Within The Skyscraper Complex - Residential Luxury Penthouses Apartments Planned, Above the Eightieth Floor, Panoramic Views - Above the Commercial Tapering, 400 Square Footprint, Skyscraper Point of Viability

400 Foot Square, Twin Observation Decks, were also planned for - Surmounting the Twin Obelisks, With Unique and Sprawling, Public Spaces - Devised to Become an Income Producing, Tourist Attractors - With Twin, Glass Pyramid Lighthouse Towers, The Centerpiece, of the Twin Plazas - These At Night, Could Be Illuminated, During The Day, An Interior, Public Fine Art Gallery, Memorial Exhibition Space, Intended - The Centralized, Elevator Bunker Construction, Was Designed, Tethering Back Rows, Of Cubed Steel Structures, Offset To Each Other in, Ascending Order, Like Staggered Masonry, of Ziggurat Architecture - An External Beam System, of Triangulated Intersecting, Cross Beams, Tied Back and Attached, to The Staggered Interior Cube Structure, Steel System - Additional Plans Were Intended, To Nickel Enhance, the Steel Foundry Formula Content, To the Hardest, Non-Flexing Strength, Possible - Hot Bolt Riveting, instead of Bolting, Was Also, Structurally Advised, To Replicate the Excellent Stability, Structural Characteristics, Of the Empire State Building - Discussed Before

The Overall Effect, Was To Be Similar, to an Glass Ant Farm, Human Movement, Throughout The Structure, Visible From Ground Level - Unobstructed Views, from Multiple Vantage Points, Along the Twin Towers, Would be a, People Watchers Panacea Solar Exposure, Would Serve To Green The Building, Requiring Less Energy, For Interior Heating - Multiple Points of Architectural Interest, Would Make, The Twin Skyscraper, a Worldwide Tourist Attraction, With Planned Accommodations, For Tremendous Tourist Traffic - Enough Exterior Space, For Major Cultural Events, Music Concerts, Outdoor Festivals, Increasing Real Estate, Income Generation, Per Square Footage - Convention Center, Built Above Subway Infrastructure, Planned for Twin Foundation Basement, Linking To Local Hotel Development, By Subterranean, Moving Walkway, Spiral Radiating, Tunnel System, Radiating from the Giant Multistory Underground Convention Center, Maximizing Income Production, of the Entire Space

Instead The Design Was Stolen, Cheapened, The Twin Obelisk Designed Merged, Into Wind Swaying Nightmare - Repeated Again, in The Scethlia Shard Building, Doubling The Dimensions, of the Central Elevator Bunker, To Inefficiently Reduce Sway - Eliminating Huge Percentiles, And Proportion, of Usable Interior Space, Redundant Eyesore, To The Monolithic Monstrosity, of WTC, Building One - Cobbled Together From The Twin Obelisk Design, Into a Fractal Fusion, of the Two Towers, Placing Undue Stress, On the Ground Fall Atrium, No Sway Arresting Components, All Eliminated, CHEAPENED, in it's Plagiarized Overall Design - Completely Unoriginal And Counterfeit I Am Embarrassed, That They Were Too STUPID - To Realize, Every Element of the Design, Had an Overall Engineering Purpose - Realizing Too Late, They Committed The Second Sin, Size Doubling The Central Elevator Bunker Shaft, of the Skethlia Shard Building - Foregoing Dynamic Twin Isotonic Obelisk Design, Reducing Interior Architectural Space, To Diminishing Point, of Ergonomic Dysfunction - Dubiously Naming, It The Skethlia, "The Fractured 'Shard', of the Throne of God" Cast Into The Void, of the Primordial Universe, To Thence Create Matter, Around Itself - Such an Architectural Design, An Oxymoron, Apparent **Blasphemy**, Made In Kabbalah Utterance, TO NAME A CRIMINAL ACT, OF ARCHITECTURAL PLAGIARISM - I Doubt The "Skethlia Shard", of "The Throne of God", Could Be Possibly, What CSUK, is

Actually, Referring Too - Such Prevarication and Hubris, on Their Part - Seems Parr, For Their NON-Architectural Course - Certainly Their Plagiaristic Corporate Tendencies, ARE CERTAINLY BIBLICAL, IN REDUNDANT EPIC PROPORTION, Blundering the Original Skyscraper Architecture Design, Not Once, But TWICE, On The River, Thames

Likewise, I Will Explain, The **MYSTERIOUS** Design Origins, of the MTA, "Trans Hub" - Originally Intended, As Was Informed, To House, a Civil War, Union Ship, (Actually Confederate Submarine, "HL Hunley") Build Near The Present, Brooklyn Naval Yards, (Actually Charleston, NC) Salvage Recovered, from The Bottom of Charleston Harbor, Where it was Sunk, in Naval Battle - It Was To Be, Maritime Nautical Museum, Anatomically Designed, From the Skeletal System, of a Sun Bleached Remains, of "Leviathan Whale" Museum Built, Within The Brooklyn Navy Yard, To Permanently House, Archeologically Salvaged Naval Artifact, The Dynamic Design Concept, Having The "Monitor Type" Iron Clad Prototype First Submarine, Exhibited in the: "Belly of The Whale" - Years Past, No Word, on the Fate of The Design, Until Passing The World Trade Center, Construction Zone - I Saw an Architectural Rendering Of It, on the Plywood Barrier Perimeter - This Was Well Before, The Secret Plans, For Building One, Were Finally Disclosed - Knew The Structure Well, Whale Bone Matrix, Was NO Accident, Of Design Coincidence - Began to Research, All My BHRA Corporate Attorneys, From The Early 2000's, Who Were Now Associated, With The Port Authority, The Transit Authority, The Lower Manhattan Development Corporation, Without My Knowledge, In Clear Conflict of Interest - Thinking Me, Out Of State, Incorrectly - They Began, The Design and Construction Process, Unawares I Had Seen, My Original Design Work Displayed - Peter Brightbill, Esquire, The Worst Offender, of THEM ALL - Soliciting Me, in the Legal Offices, of Holland and Knight, In Clear Earshot of Reliable, Cooperating Witnesses, in 2001-2 - Then Assuming Design Credit, "Shyster Lawyer, Turned Skyscraper Architect", Pathetic Excuse, For A Human Being - I Hereby Guarantee - This Truth Will Prevail, Brightbill, Bergen, Papert, WILL DEFINITELY CRACK UNDER PRESSURE - Like The Bastardized Architecture They Stole From Me - They Are, Ignorant Shyster/Political Hack Lawyers, NOT ARCHITECTURAL ENGINEERS: At Least Robber Baron, P. Cooper, INVENTED, "Tom Thumb" Locomotive

The Architects, Who Stepped In, To Adulterate The Work - Are No Better, In My Assessment - To Take Another's Plans, And Affix Your Professional Stamp To It, Akin To, Cemetery Grave Robbing, Disguised as Legitimate Archeology: Precious Artifacts, Stolen in the Night, Blithely Unconcerned, With Their Cultural Origin Provenance, Nor Actual Legal Ownership - Clearly I Am Sickened, By Such Puppet Men: Santiago "Chile", and "Player" Piano - How Many Sheckles of Silver, Did It Cost You, To Sell Your Souls, To Silverstein - Buy a Judas Traitor Tree, and Dangle From It, Or Hang From Your Own, Siege Engine Petard: Till The Crows Pluck Out, Your "Copyist" Eyeballs, As In The Days Of Old - Read This And Weep - I Have Regained My Dignity, Retrieved My Authorship, Your Architectural Plagiarism, Crime Mounted Upon a Crime - I Am Sure, You Will Say, It Was All, A, "Big Misunderstanding", Phenomenal Coincidence - I Doubt That Rhetoric, Has ANY Currency, OUTSIDE OF HELL - Projects Of This Magnitude,

Have Extensive Crumb Trails, Connecting Forensic Evidence - Your Plagiarism Will Be Reckoned With - Your Professional Name, Tarnished Beyond Repair - I Hope It All, Was Well Worth It, As The Snail Trail Slithers, In Your General Direction - To Your Guilt Soiled Bandit Doorstep Investigative Reporters Love, Scandalous and Juicy Story to Tell, And You're It - Have Now Provided, The Ways And Means, For You To Sink and Wallow in It With Audacity To Lie, About HL Hunley, Confederate Submarine, Being Union Iron Clad Displaced Historical Naval Archeology, Planning Unaware Yankee Claim Jump Museum

The Coney Island Amusement Park, Master Plan, Was Likewise Highjacked, and Duly Plagiarized, by Mayor Bloomberg, and Dan Doctoroff - Palmed off, to the Coney Island, EDC, in the "Raider Spirit", of Robert Moses - Sanborn Architectural Illustration, Orthoimagery Projection, Exhibits, of the Super Coaster, Year Around Amusements, Future Hotel Development, Proposed Restored, Historic Iron Pier, and Expanded Boardwalk Attractions, Urban Design, Intermodal Connections, by Land and Sea, Here Included, Major Collaboration Effort, of BHRA Transportation Expert/Design Team, My Own Focus Being, the Amusement Park Design Itself, Set around the Landmarked Historical Permanent Attractions - Solicited by The Former Borough President, Marty Markowitz, On Official Letter Head Stationary - For BHRA/DIG, To Submit, The Master Plan Design Work - Once Again, the Entire Concept Stolen, No Credit, or Financial Renumeration, Of ANY KIND, Offered or Given - In Fact, initially Refused Admission, as our Master Plans, Were Exhibited, In Coney Island, By CIEDC - Within a Solitary Year, Groundbreaking Ceremony, Small Portion, of Our Original Supercoaster Concept, "The Thunderbolt" Construction Began, Indicated Stolen Design, in the Illustration Appendix

Robber Baron, Corruption Levels, Such As This, Not Seen Since Tammany Hall, I Suspect, A Whole New Level, Of Undue Influence, NEVER Seen Before - All These Occurrences, During the Auspices, of The Twelve Year, Bloomberg Mayoralty - Multiplying His Billionaire Fortune Tenfold, He Destroyed This City, I Was Born Native The Conceptual Design Renderings, My Company Produced, Were Directly Used, by The Sanborn Map Company, Producing Architectural Renderings, PAID FOR BY MAYOR BLOOMBERG - There is NO Architect of Record, The BHRA/DIG Design Theft, Went Direct To Sanborn Map Company, The Illustration Plates Expedited, At Great Cost, Were Rendered, From Our Master Plan Design Concepts - Published in the Major New York Press, Our Names and Company, Conspicuously ABSENT FROM IT -
Six Months of Urban Planning and Amusement Park Design, Notice Left Uncredited -
I Also Hear The Former Mayor, is a Frustrated Night School Engineer - Who is Phobically, Afraid To Drive Cars - Is Obsessed With The Dubious "Rhodes Scholar" Legacy, of Robert Moses, and "The Powerbrokers" The LaGuardia Administration, Depression Era Crooks and Vandals - Why Pay More, When You Can, STEAL EVERYTHING, THAT IS NOT NAILED DOWN: I am sure All Those Designer, Cocaine Based Drugs, Mike has Been Smoking, FOR YEARS, will Be Catching Up With Him: Sooner Than Later, Good Riddance, Mayor Bloomberg, "You Are, A Legend, In Your Own Mind" - Mental Midget We Also Owe, An Additional Debt Of Credit, to 'Little Mikey', for Destroying a Federally Funded, ISTEA Enhancement, BHRA, Historical Trolley Project, in Redhook, Brooklyn - An Entire Fleet, of Seventeen Historical Trollies,

Scrapped, or Unlawfully Donated, to Other Regional Train Museums, Without Legal Notice - Our Valuable Paperwork, and Franchises, to Build a Trolley System, in The NYC Streets, The First One Given, in OVER Eighty Years - Revoked by Bloomberg, Through NYDOT - The Atlantic Avenue Tunnel, Discovered by BHRA Chairmen, Founder, Robert Diamond, in 1980 - Also Closed Down, by New York City, Through the FDNY, and NYDOT, in a Real Estate Power Grab - We Have Been Fighting, in the New York State, And Federal Court System, Ever Since - Apparently, This Is, The Signature Brand, of EXTREME Ingratitude, This Napoleonic Sychophant, MIDGET, WAS In The Habit, of Distributing, for Services Rendered - Fortunately, At This Juncture, The New City Government - IS CLEANING OUT, HIS REMAINING PIG STY - Some Day Soon, A Bronze Statue, Will Immortalize, Your Foulest, Political Legacy - Dedication Plaque, Eternally Reading Like Epitaph: "We REALLY Hated You Mike, When You LEFT OFFICE, Everyone Checked Their Wallets" - Go Caribbean, During Hurricane Season

'Living a Life of Quiet Desperation' Is Untenable In These Times, We Are Reduced To Deers, in the Headlights, Waiting To Become Roadkill - I Have Tried in Vain, to Remain Apolitical, in My Philosophical Existence, to No Avail - The Promethean Atlas Giant, in My Ancestral, Ancient Boernician, Highlander Soul, My Hereditary Basque Animosity, for Imposed Imperial Rulership - Thus, I Vent My Spleen, I Slay The Olympian Eagle, of Imperious Zeus, Into Auguries, of VIVISECTED TRUTH - No Longer Silenced, Chain Unbound,  Self Actualized NOW - Beyond The Fear, That Restricts Free Philosophical Discourse: I Will Tell It, As I See Fit, To Disclose, With Clarion Calling, The Suppressed Secrets, of My Creative Legacy, Active Contrition and Spiritual Atonement, Regaining The Center, of Philosophical Wisdom, IN FULLEST DISCLOSURE  - Call it if You Will, An Artists Manifesto, Poets Dream, Righteous Invocation, To Those Who Have Wronged Me, I Catalog Their Crimes, Thus May They Live in Infamy - Dantesque Inferno Cantos, I Have Reserved, For ALL Of Them - Immortalized, in Genius Formulated Literature, I Shall Assure, Their DUBIOUS Notoriety, IN LOWEST HELL:

The Sins of the Fathers, Legacy Unto Their Tyrannical Sons, for GENERATIONS, Yet To Come - The Rotten Fruit Falls Closest, To The Thunder Struck Tree: That This Book INTENDS, as Razor Sharpened, Woodsmen's Axe, Will Take Pleasure, In Tree Felling, Into Witch's Pyre, Accursed Wood Kindling: Religious Instruction Gleaned, by Malleus Maleficarum - One Must NEVER Suffer a Witches CABAL, To Live Impiously UNTRAMMELED - Their Names, Now Invoked, Their Wrongs, Transparently Displayed They Will Never Again, Trespass Against Me: I WILL Testify Thus, Upon This Thing, In VERITAS - Let Freedom RING, Liberating Verity, From Apocalyptic Corruption, Exposure Now Begins, Unfolding Before Them - No Quarter Given, No Mercy Shown, No Pact With The Devil, I Rebuke And Hate - Legion, Are His Minions, In This Declining State - Diatribe Worthy, Stated With Cold Precision, Genius Logician, With Incisive Mind

I Will Not Retract, Or Revoke, a Single Word Of It - Truth Will Conquer ALL, The Fallacy Of Their Conspiracy, Herein Told: Beyond Rage, Personal Offense, Martyrs Retribution, They Have Hardened My Heart - Good Will, Now Exploited, Inverted Now, To Inveterate Hatred - To Articulate, The Misery, I Have Suffered Greatly, At Their Purloining Hands -

I Revel, In The Opportunity, To Expose Them NAKED, Before You Now - The Silence Has Been Broken, My Conscience Clear, My Goal and Purpose Completed, Completely Rendered, and Set Down Here, Satisfying My Philosophical Need, to Finally Establish Architectural Skyscraper Design Provenance, Exegesis, to My Epic Poem References - No Longer Will I Tolerate, Their Blighting Influence, Remaining Free and Clear - Many Have Been Their Victims, Let The Populace Come Forth - I Encourage You, Protest Unanimously, Vigorously Assert, Your Inalienable Rights - Do So Robustly, Before Your Freedoms, Are Taken AWAY From You - The Founding Fathers, of This Country, Are Turning, In Their Graves - The Genius, of Their Political Conception, Teetering, on the Brink of Extinction My Case, Just One Example, Of The New Regime, of Oligarchic Orthodoxy - Brought Back, To Tributary Mercantilistic Colonialism, of Multi-National, Corporate Interests: Individuals Subjected, By The Oppressor State, Reduced to Utilitarian Cogs, Means of Production, In Diabolical Mincing Machine - Fortunately I Have Survived My Trial, Regained My Own Path Again, I Leave This Living Record For You, To Further Reflect Upon, Calculate The Ramifications, It May Have, In Your Own Life - Particularly, To Fellow Intellectual Workers, Prodigy Savants, Creative Geniuses, Inventive Artists, Encyclopedic Minds, Grand Master Generalist Polymaths, Again I Salute You All! - Cast Off The Yoke of Tyranny - You Are Not BORN ENSLAVED, Nor Subject, To ANYONE: Use Focused Genius, To Ensure Freedom, Especially Your Own - If My Chronicle, Has Awakened, This Intrinsic Philosophy, Within You, I Have Accomplished Much - Rage Against Anything Oppressive, That Would Curtail, Your Personal Destiny, The Creative Impulse, Most Indomitable, In The Most Brilliant Minds -

Let Not The Poem Fire, Be Ever Extinguished Within You, Wisdom Teaching Continue, Around The Ancestral Gathering Places - Abusers Are Emboldened, By Revisited Abuse, Victimizing The Victim: Stand For None Of This: Lest, I Be Hypocrite, To My Own Philosophy, Having Written Here, To Do Exactly That: Like The Bible Prophets and Patriarchs, Placing A Curse Upon Them, So That They And Theirs, Shall Go The Way of The Dinosaur - Such Things Go Not Unavenged, If One Have The Courage, To Bare Witness To It, Speak Righteously Against The Wicked, Protect the Wisdom, Within Your Heart - To Be Offered Up, Not As Martyrdom Sacrifice, But Through Works and Deeds, Of Personal Creativity, and Autonomous Faith Worship - Freed From Bondage, Condescending Autocracy, Compelled Not, Against Ones Will, To Forego Individuating Chosen Path - Squander and Rob, Not The Labor of Others, Do Not Take, or Steal, What Is Not Yours - For In Time, Truth Be Known, Your Name Be Mentioned, People Will Spit, Upon The Ground - To Covet, The Property Of Another, Deprive Them Of Their Due, Eventually There Comes, A Time - Where Everything, Will Be Taken From You - Often Raised to Dazzling Heights, For All To Envy, In Vainglorious Prominence, Then Smote Down, Crippled, With Exposed Feet Of Clay, To Crawl Like a Poisonous Serpent, Stomach Dragged, Upon the Floor, Anathema Then Ostracized - Made Example Of, To Those Who Would Trespass, With Legs Still Attached - The Wages of Sin, IS DEATH - Be It Physical, Spiritual, Intellectual, It Matters Not - To Challenge Fate, With Hubris, Incite the Wrath of God, Thus Engender Damnation, Unto Yourself - Facilitate Undoing Divine Grace, Has Given Me Intellection, Into Mysteries, Few Have Ever Known -

Gathered Thus, Empirically Evidenced, Giving These Words Freely, Unto The Wise Among You - Humble Yourself Before the Lord, Make Peace With Your Creator, Let His Creativity Flow Through You, Like a Niagara of Discerning Energy, Let Intuition Be Your Guide and Savior, Against The Supreme Arrogance of Hubris, The Wisdomatic, Must Ultimately Strive, Against All Odds, They Shall Prevail - Might Makes Right, a Fallacy, Time Bares This Out - With Mantle and Stave, in Prophet Hand, I Will Not Retreat, Battlefield, In Thick of Combat - Careering Forth, Placing My Existence, Unto The Will of Fate - A Philosophical Life, is Often Challenged, By Such Trials, of Fire and Ordeal, Crusading Onward, as My Warrior Ancestors, Before Me - To Find and Slay, The Dragon Speaking To The Stones, The Labyrinthine Lair, is Found - To Do No Less, Is The Heroes Destiny: Careering Through, The Valley of Death, Will Fear No Evil, No Retreat

Armoric Knight, Protected With Heaven Wrought Metal, Impregnable, I Sally Forth Some Are Born To This, It Is The Mysterious, Fate of Things, I Sojourn Never Looking Back - Thus I Seek The Meaning, Of My Existence, In The Final Destination Exodus - Let No Man, Obstruct Me From My Chosen Path - My Wrath Knows No Bounds, When Raised To Fury - Consider This Well, In Your Own Magnificent Journey - Anything Less Than This, Is Unacceptable - I Will Leave This Life, Leaving No Regrets, Behind - Suffice It To Say, I Lay My Philosophy Down, Pick Up The Claymore, Of Clan Lamont - There is Time For Peace, and Time For War - I Would Bring Such War, To The Enemy, Sparing No One, My Cleaving Sword - Interpret This As You Will, Be It Reality or Metaphor - Every Man a Hero, in His Own Story - My Life, In Part An Enigma, Secret Mythology, Folded Conundrum, Cannot Readily Disclose - Rest Assured, I Have Confronted Much, Seen Fantastic Things, Only Whispered Of, Obscured In Legend: My Eyes Grown Weary, From Discerning, The Most Wicked Things Imaginable - Yet Still, I Battle Onward, For It Is The Only Way, Homeward - No Other Strategy, Has Been Availed To Me: Dragon Slayer Avenger, This Is My Righteous Destiny - Champion of My Own Truth, Authoring My Own Philosophy, Sealed in Blood Covenant, To Achieve Prophetic Purpose - Answering To No One, I Am Born In Freedom - Those Who Trespass Against Me, and What Is Mine, Have Known Transcendent Wrath You Are Hereby Forewarned, Do Not Take Me Lightly, Or Mistake My Ultimate Purpose, Watch, As The Scandalous Spectacle, Revealeth Itself, As Unjust Power Grovels, Prostrate, Falsely Repentant, Mud Crawling, Like Pregnant Swine Herd, Upon The Ground - These Prophesied Things, Are Not Of This World, I Would Leave It, At That - 'There Is More In Heaven And Hell, Than Dreamed of, in Man's Philosophy' -

I Will Attest To This, Abiding By Natural Law, The Phenomena of Miraculous Things, In Due Time Becometh Rarified, Transgressions Are Duly Rectified, Beyond Human Comprehension, Or Corrupting Breach - Inspired As I Am, To Write This Testimony, I Know The Process Described, Has Now Irreversibly Begun - Poisoned Fruit Ripened, Upon The Stagnant Vine, For Horticulture Scythe Cleaving - Righteous Words Have Power, To Change The Way of Things, The Reality Paradigm Shifts, Renaissance Begins - The Time of the Tyrant, the Oligarch, the Autocrat, Coming To Conclusive End Archaic and Obsolete, Burdensome Upon The World, Becoming Conspicuously, Irrelevant, Unproductive Detritus Of The Old Dying World - The Intellectual Worker of

Genius, Must Be Accorded His Rightful Place - Conceiving in Righteous Solitude, The
Individuating Laboratory, of Hermetically Sealed, Imagination, An Original Universe, Self
Created - Sacred Autonomous Space, Sanctified - From Which All Things, Masterpieces
Of Intrinsic Worth, Are Creation Born - It Is A Lone Frontier, Beyond The Outer
Boundary, Beyond What Has Ever Been Before - The Realm of Signifying Dreams,
Thus Meant To Enter Alone - Solitary Journey, Fraught With Dangers Manifold: No Map
To Guide, Where the Strange Land, Remains Primordial, Wildness Unsurveyed - Only
The Interior Compass, Becomes Intuitive, Dead Reckoning Guide, Series Of Halting
Steps, Along the Expeditionary Way - Leads One, With Uncertainty, To The Desired
Goal - Envisioned Innovation, If Tireless Will, and Capricious Fate, Would Have It So -
My Mind Has Transversed, Such Unforeseen Lands - A Biopic Sojourn of a Million Miles
The Kindred Creative Spirits, That Have Gone Before Me - To Pass Their Memorialized
Landmarks, Along the Labyrinthine Way - Others I See, As Lost Spirits, Phantoms, Still
Wandering, Lost In Circuitous Speculation, Made Self Mad, As If In Eternal Haze - Only
Those With Compelling Ideas, Should Cross, The Borderland Portal - Like Burning Star
Illuminating, The Canopy of Night - The Idea Itself, Will Guide You, Above The Dismal
Fray - An Irresistible Force of Nature, Compels You Forward - An Unquenchable Thirst,
For Original knowledge, Not Yet Found - Into This Mundane World, Hunger Insatiable,
Finding Common Knowledge, Unpalatable - Unable to be Quenched or Slacked, by the
Pervading Order, of Unoriginal Tedium - To Return, Like Jason and the Argonauts,
Golden Ram Fleece, Taken From Golden Bough - Like Saintly Mantle, of the Prophet's
Inheritance, Passed Down, In Righteous Meritocracy, To Prophesy and Faith Heal -

To Kill The Messenger, Is The Greatest Sin, In Damnation, Mankind Will Wander
Aimless, Do Not Silence, Constrain, Nor Abuse Him, Lest Ye Be Thunder Struck:
Wrathful, Unimaginable, Annihilating Power: By Grace Protected, He Speaks For ALL
Though Time Moves On, The Ages Pass, These Cardinal Laws, Reman Everlasting
Though More Subtle, In Modern Times, These Things, Come To Pass, Though Less
Apparent - Such Phenomena, Closely Monitored, To Prevent General Hysteria
Ignorance Is Not Bliss, In The Final, Analysis, Suppressing the Apparatus, of Course
Correction, Infantilizes The Uninitiated - Serves Only To Augment and Magnify,
The Wrongs Committed, Maintains the Most Wicked, in the Highest Places - Until There
Can Be Found, No Honest Man - Though Illuminating Magic Lantern, Shows The Way -
What Then, When Elohim, The Lords of Heaven Arrive, Repulsed By Sodomites,
Who Would Prefer Worse Rape - Leveling Twin Cities, Unredeemable In Decadence,
No Stone Left, Standing Atop Another, Shadow Figures Left, Eviscerated Humanity,
Form Nuclear Blast Basin, Lifeless Deadened Sea Arises - Warning The Prophet and
His Two Daughters - Disobedient Wife, Forsaken, Turned Around, To Pillar of Salt - To
Seek the High Ground, of Mountain Cave Grotto - For The Strafing Rage, of The Lords
Of Heaven, Is Soon To Be Unleashed, Upon The Dishonest and The Unrepentant

Oftentimes, I Speak, in Parables and Riddles, in Keeping, With My Eclectic Nature -
Bouncing from Subject to Subject, as it Pleases Me - A Fortean String, For You To
Connect - I Write, As The Spirit Moves Me, That Is All - I Am Not A Literary Entertainer,
Pundit, Sarcastic Wit - Philosophy and Epic Poetry, Subjects, of Avocational Interest

This Biopic, Was Thus Inspired, Outside My Usual Comfort Zone, of Expertise - Architectural Skyscraper Design, One Of Many Things, I Have Been Involved In Having Been Trained, as a Fine Artist, Saw Opportunity, To Express My General Aesthetic Views - And Make My Position, Crystal Clear, Upon the Criminality of Architectural Plagiarism - My Own Experience in This Area, Makes Me Uniquely Qualified, To PONTIFICATE, Upon It: I Hope It Has Served It's Purpose Well, To Dissuade The UNCREATIVE, NEVER Steal, From The Creative Artists, Sinful:

As in the Norse and Greek Mythologies, Often a God, Disguised as a Beggar, Would Come To Plead, For Simple Food and Humble Shelter - To Test and Divine, The Nature Of Their Host - To Ascertain, What Type of Human Being, Would Welcome, or Turn Away, a Person in Dire Need - To Address, Their Specious Nature, With Olympian Wrath, or Reward Them, for Their Humanity, One Thousandfold - With Cornucopia By My Reckoning, Often Accurate and True, I Would Say Such Days, Are Returning - As Man's Inhumanity to Man, Continues To Increase Manifold - It Will Draw Down The Wrath, Of The Higher Power, Call It What You Will, Be It Pure Energy, Transcendent Being, Anthropomorphized - Essence Ethereal, Like Unnameable Tao - Has The Power, To Create or Destroy: Greatest Creator of First Genesis: It Protects, The Creative Worker, Generative Forces of Nature, Kindred Emulation, of Righteous Worship, Thus Sanctifies and Guards, It's Own Kind: The Is The Way of Heaven, The Path of Nature

The Most High God, Conferred Revelation, The Monotheistic Nature of Himself, Upon The Scethian, Indo-Aryan Priest King, The Melchizedek of Salem: A Shard of the Throne of God, The Skethlia, Ruler of The Nomadic Peoples, of the Caucus Mountains, And Black Sea Strongholds - Ancient Sogdiana, A Name and Place, Associated With the Moon Goddess, Diana, The Zoroastrian and Mithraic, Eternal Fire Religions - The Royal Seat, the Caucus Plateau Pastural Basin, Of Royal Skythians/Scethlians/Sogdians: Proto-Iranian Metaphysical Culture, Magician-Supermen - Later Through Instructional Transmission, of Said Revelation, to Babylonian Prophet, Abraham, The Shard of Throne, Was Conferred, Genetically Transferred, Through Original White Aristocracy, Of Western Europe, Through the Moravingian, and Magdelanean, Royal Blood Lines - Known For Extraordinary Hair Length, Crimson Hued, a Genetic Marker, That Indicated Their Metaphysical Strength, Primogeniture Sovereignty - Talking To The Stones, The Fractal Shards, From The Throne of God, Descended From Elohim, Geographically Isolated and Segregated, Holy Mountain Ranges - Reincarnates Aware, Of Their Unbroken Mystical Dynasty - Tracing Back, to the Reign of Melchizedek, Priest King, of The Salemites - The Throne of Scone, The Stone of Destiny, Indicates a Gathering Place and Time, For This, "Missing Tribe", The Latter Day Saints, Translating From Spirit, Back Into Physical Being - Asserting Rightful Heirdom, to Their Kingdom - MALKUTH, Left Behind - Like Camelot, Thule, Shangra-La: Inter-Dimensional Nature, Set Upon Scethian/Skethlian, Mountain Summits - Destined To Appear, The Fractured Shard, Rejoined To Throne, of Heaven - An Obscured Prophesy, That Should Be Known Like Celtic Psychopomp God, "Ognios", **Sharpened Rock**, Ornate, of Hermetic Speech: The Black Aristocracies, of Modern Europe, In Influential Decline, Are Aware of This - Beginning, With the Royal Pretender Imposition, Carolingian Empire - The Royal Blood

Plantagenet's, By Windsor Tudor's, Dethroned - An Unstable History of Rulership, WithWhite Aristocracy Purgation - Lost Royal European Dynasty, Hunted Down Systematically, For A Thousand Years - It Grieves Me To Know This, As Royal Boernician Scot - It PERFECTLY Explains, the Machinations, Detailed in This Book - Above and Beyond, Common Persecution - Extrapolated in Theory, of Forensic Motivation, It Speaks Much, To Blasphemous, Cult Activity, Scottish Clan Campbell @ "Center of Mass" (Lamont Massacre, at Dunoon, c.1646) - Naming My Stolen Design, The "Skethlia Shard", is Not An Arbitrary Accident - The Metaphysical Reference, is Specific, Arcane, Obscure, "With Eyes Wide Shut" - The Ancient Legends, of These Details, Are Readily Researchable - The Motivations of These Plagiarists, Still Patently Unclear - Perhaps Meant To Be, Bringing Forth, My Fascinating Story - Only Time Will Tell, If My Theory Has Fathomed, Their Convoluted, Ulterior Clan/Cult Motives - By Trespassing Upon Me, They Have Tipped Their Hand - I Have Learned From The Experience, Putting The: "Shard Pieces Together" In, "Speaking To The Stones" - Attacked As I Was, Aroused and Awakened, Become Self Aware, Of Who and What I Am: Well Worth The Lost Building Design, Clan Campbell's, Cult Sabbateans Beware - Destruction of a Business, Through Secret Cabal, of Surrounded Criminal Conspiracy - Their Catalog of Crimes, Led Me, To Do The Research, Was Destined to Completely Apprehend - Made Self Aware, With Total Recollection, Things Past, Present and Future It Does Not Bode Well, For Those Who Wronged Me - A Paradigm Shift, in World Philosophy, Is On The Event Horizon, Making People, Such As These, Obsolete, Redundant, Culturally Irrelevant - They Are Prophetically Marked and Spiritually Doomed - They Have Warred Against Me, Mine Eyes Have Thus, Seen Their Fate

The Skethlia-Shard, is Not a Stone, But Aristocratic Dynasty, Reincarnation Line, Of Royal Related People - "Speaking To The Stones" To Speak To Ones Own Ancestors, Coronation Upon The Throne of Destiny - The Shard, is a Psychological Concept, Of Genealogy Conjoined, Yet Separated Race - Psychically Linked, Through Reincarnation, Separated By Time and Space, To Protect Themselves, From Attempted Adversarial Annihilation - They Will Appear and Rise Again, Breaching Time Itself, Rejoining, The Fragments of the Shard, Congregating Back Together, To Repair and Complete, Become Throne of Heaven - "Thy Kingdom Come, Thy Will Be Done, on Earth as in Heaven" Is the MALKUTH-Kingdom, A Living Community, of Reincarnating Beings, Gathering Together, into Lamed Vav, The "Thirty Six" Righteous, Variants of Human Astrology, Forming Holistic Homeostasis, Upon The Planet, The Mighty Men of Biblical Testament - Sent Forth in Pairs, ELOHIM Watchers, The Lords of Heaven, Come Again Charismatic Gestalt, Synchronicity Field, of Soulmated Pairings, That Liberate The World in New Age - It Has Nothing To Do, With a Legendary Rock, Or A Stolen Building Design, Named After It - Those Are Simply Misconstrued Prophetic Signs, And Relativistic Metaphors, Code Broken, By The Initiated - The Structure in Question, is a Human Community, Not a Lifeless Temporal Thing - Apparently the Wicked, are Always Outwitting Themselves - I Have Come To Observe, This Vile Tendency, Destructive In Nature, They Ultimately, Unwittingly Serve Goodness - While

Revealing Themselves, As Anathema, In The Machination of Their Vilest Process -
Transgressors, Upon The Nature Way Of Things, To Serve As Fertilizing Dung -
Trodden Under Foot, in Protesting Rebuke, So That Paradise Gardens, May Grow
Above Them, Their Unmarked, Unhallowed, Shallow Tombs - Such Is The Nature Of
Things, As Well It Should Be, The Wicked Made Thus Low, To Suffer Anomy In Death

The Metaphysical Knowledge Provided, By This Plagiaristic Odyssey, Well Worth
The Pain It Has Cost Me, Greatly - With Philosophical Repose, Triumphant, in the
Realm of Ideas - As They Glory, in the Tainted Abortion, of it's Horrendous Counterfeit -
Made Fortunate, To Have Created a Book, To Publish It, To Leave Detailed Provence
and Intellectual Record, of Inspired Conception: Knowing The Pleasure of Original
Creativity, Not The Loathsome Servile Sloth, of the Technician Copyist - Ideas Continue
To Flow, Taking Polymath, Metamorphosis Form, Leaving Those Who Would Steal From
Me, In Cloud Wake, of My Wandering Nomadic Dust - Doubting They Know Will Where
To Steal, Or How To Find Me, I Am More Resourceful, Original, and Much More Clever
Than They - I Have Taken Back, What Has Been Stolen From Me, Hermetically
Enclosed It, In This Book - The Idea, The Most Important Thing, After All - What Was
Once Art, Now Become Artful Science - Giving Credibility, To Mystical Superstition -
Would Tutor Them, On Errant Slew, of Misconceptions - Sent To School Them, Or Send
Them Packing, It Is Irrelevant, In The End - Redemption of the Wicked, is Rarified
Miracle - Even Prophets, Often Can Find, No Honest Man, With Magic Lantern -
Decadence and Debauch, The End Of Days, a Self Fulfilling Prophesy, They Covet
Much - Requiring What I Have, I Have No Use For Them, I Turn Instead, To Greener
Pastures - Gang Of Thieves, They Are Bound Together - Their Crimes Compel Them, To
Oath Secrecy - Like Scorpions in a Sealed Jar, They War Against Each Other, Endlessly
No Peace Possible, In Such Self Imposed Hell: Wherever They May Congregate -
Hoping This Book Agitates Them Further, To Shake Bell Jar Most Vigorously, As They
War and Conspire, Within It's Precincts - Scorpions, When Aroused, Often Strike Their
Own Head, With Poisonous Tail, Enraged In Berserking, Suicidal Tendencies -
Suspecting, The Creator, Had This Intention In Mind - To Exhibit The Natural Pitfalls,
Of Such Insect Like, Self Destruction, Unproductive Behavior - Human Emulation, of
The Scorpion Insect, Exists Beyond The Point, Any Possible Method, of Schooling -

Was It Not, The Paid Sophists, The Guild Of Lawyers, Of Ancient Athenian Polis -
Who Brought Spurious Charges, Against Dialectician Philosopher, Socrates, The
Espouser Of Uncomfortable Truths: "For Corrupting The Morals Of The Youth" Who
Sought His Wisdom Out, Freely Given, Nothing More - The Hated "Gadfly of Athens,
Of The Agora Marketplace" Made Into an Open Public Forum, On Daily Basis:
To Eviscerate and Rebuke, Their Complex Method, of "Sophisticated" Lying -
Mesmerizing Targeted Populace - With Intellectual Confusion, and Will Breaking
Acquiescence - Socrates, Sought To Course Correct, This Propagandist Mutation
Of the Declarative Reality, Based in Empirical Efficacy, of the Sophist Abrogated
Language - To The Original Golden Standard, of Sensorial, Philosophical, Linguistic
Understanding, from Obscured Sophist Rhetoric - For His Free Speech, and Public
Protest, Facing Dreaded Hemlock Goblet, of State Execution, Or Political Exile, Ten

Year Ostracism, in Advanced Age, an Alternative Death Sentence - From Beloved City State, Democracy of Athens, Would Not Leave - Choosing the Latter, Choosing the Poison Draught, To Murder Himself, To Bring Permanent Silence, "To Gadfly" - Killing The Philosopher, But Not The Philosophy - Horangued As I Was, By Paid Sophist Agents, Of My Own, Surrounded - I Have No Intention, To Entertain Their Solomonic Judgement - Martyrdom to Me, Is Philosophically Untenable - My Guiding Daemon, My Inner Locus of Control, Unique Universe Unto Itself, Requiring No Specific Place or Time, To Prosper, The Mind A Kingdom Unto Itself - Where Philosophy Empowers, The Philosopher King - This Surely Is No Athens, I Am Not A Socrates, Near The End Of Years, In Fact In My Prime - Yet My Persecutors Have Excelled, Their Ancient Predecessors, Innovating Forms of Modern Sophist Tyranny - I Will Spit Out The Hemlock Goblet, Upon Their Specious Faces - No Sacrificial Fowl Offering, to Asclepius I Will Not Be Exiled, By Lesser Men, Or Driven Out, In Exile, To Live Among Barbarians, Ostracized Against My Will, In Old Age Cast Forth, Into Unforgiving Wilderness Tempest Boernician Royal Scot, Iberian Basque, Scythian/Skethlian/Sogdian/Melchizedek, I Am Of Unbroken Barbarian, Indo-Aryan Bloodline, Fearless To Return Homeward, To Vigorous Freedom, The Living Power, of My Ancient Truth Seeking People - Turned Avenging Lions, At The Fortified Moated Gates - Citadel Walls Crumbling, Around Captivated Decadence, An Imploding Declining Civilization, The Carrion Raptor Birds, Encircle High Above - Drawn, To Growing Wafting Smell, of Fetid Death, And Impending Fear, Enervating War - Truth No Longer Spoken There, Long Forgotten Lost Dialect, Surreal Reality, of Strange Things, Spoken Of, Apocalyptic Imagination Taken Hold - Replacing Verified Language, With Incomprehensible, Beggars Canticle, Pathology Phantasmagorical, Swindlers Babble - Nothing, So Fundamentally Contrived, Born Malevolent, Untrue, Synthetically Artificial, Can Ever Long Hope, To Endure in Nature:

Born Of, The Most Ancient Barbarian People - Fearing Nothing, Beyond The Crumbling Walls, of Spent Civilization - The Barbarian Philosophy, That ANYTHING is Possible - Freedom Of the Individual, Most Sacrosanct - All Things of Righteous Purpose, Will Be Revealed, In Divine Right Timing, Everything Required, Can Be Provided By Nature - The Righteous Are Protected, The Wicked Destroyed - Baring False Witness, To Steal From Another, Coveting What Does Not Belong To You - Punishable By Trial Of Fire and Ordeal - Fate Decides Death, At Projectile Edge, Viking Bearded Axe, Singing Arrow, From Sprung Long Bow - Like The "Shard of the Skethlia", Cast Into The Universe, At Speed Of Light, Gathering All Matter Unto Itself, With Magnetic Gravitational Pulling, Accelerating Charismatic Power, Creating Parallel Throne, Exact Material Replica, From Where It Spiritually, Came From - Finding The Lost Fragments, Gathered Identical, in Metaphysical Cosmology, To Reconstruct Itself, as Lost Aristocracy Reincarnates, Chosen Dynasty Returned, to Build The Eternal Kingdom - Steal What You Will, It Means Nothing, In Retrospect, These Things of Absolute Power, You Allude To, In Mystical Reference, Far Beyond Your Knowledge, Controlling Thrall, Do Not Concern Yourself With Such Things: Like Children Playing With Matches, You Will Soon Learn, After, BURNING YOURSELF - You Were Not Chosen, To Know Such Things, Yet Another Hackneyed Idea, You Stole, In incessant Wandering - Self Deluded, Into

Belief System, Like Architectural System, Partially Comprehended, Ignorance and Spite Broken - Even The Vocabularies Involved, Are Indo-Aryan - From Dialects Spoken, Upon Anatolian Plane - Symbolism Derived, From Mithraism, Zoroastrianism, The Nordic Legends, Eastern God Names - Derivative Counterfeits, Of Warrior Religion, Nomadic Horsemen, Mounted Archers, Heroic Sea Mariners, Not a Race of Professional Victims, Liars and Slaves - The Agricultural "Scythe", The Warrior "Bow", Associated With Distinct Shape, Of Crescent Moon - Share Etymology, to the Scythians/ Scethlian/Sogdian, in the Caucus, Carpathian, Himalayan, Interlinking Mountain Ranges From Mongolian Plane Steps, To Shores, of Black Sea, From The Hunnish Hordes, Along Shores of Danube, To Melchizedek Priest King, of Salem, They Are Mountain Dwellers, Royal Shards, Destiny Projecting: "Sceth"/Arrow Heads/Ship Prows/ Sharpened Projectiles/Blind Reckoning, Internal Compasses, Roaming Nomadic Warrior People, Internal Daemon Guided, Vision Quest, Shamanic Wanderers - The Animus Mundi, of the Skethlia Stone, Original Inventors, of Western Monotheism, Genius Inspired, Well Before Written Bible, Babylonian Captivity Compilation - Converting Abraham, Of Babylonian Ur, Once Idol Worshiper, Making Graven Images, Into "Shards" In The House Of His Pagan, Cult Priest Father, Broken Against, Grunsel Edge-Door Saddle, Unclean Profaning Portico - Then Prophetically Inspired, Searching Out The Salemite, "Promised Land", of the Melchizedek Ruler, Arrival Long Expected - From Priest King Psychic, of The Black Sea, Anatolian Plane, Persian Steppe, Caucus Mountain, Vast Nomadic Indo-Aryan, Vasty Royal Empire - Syncretic Cultural/Genetic Fusion, With Mongolian/Inuit/Siberian/Caucasian, Horde Forming Alliance, Like Hun and Mongolian - Algonquin, "Manitou" Great Spirit Theology, During Bering Strait Ice Bridge, Eastern Diaspora, Colonizing Arctic Western Europe, Vast North American Continent - Dominated, By Universal and Analogous, Aboriginal Form, of Syncretic Monotheism: Since it Has Bearing on the "Name" Etymology, of Stolen Architectural Design, The Naming of it, Passing Strange - I Wanted To Share, My Knowledge of It, As Polymath Indulgence - Apparently, Some Would Be Poet, Or Pseudo-Mystic, Would Name a Building Thus - Taken From Experience, You Should Walk on Eggshells, When Mixing Religious Esoterica, With Architectural Skyscraper, Plagiarism - Someone, is Bound To Catch, the Obscure Kabbala Reference, And Under The Circumstances, Not Find It Very Amusing - Doubting Most, of London's Population, Knows the Definition, of "The Skethlia" Shard - Having Done My Utmost, To Expose, This Charade of Skyscraper Design, False Authorship, With Due Diligence Afforded Me, I Set Down The Actual Accounting, of Intellectually Conceived, Authentic Provenance - As Official Record The Original Engineering, That Attends, The Inherent Stability of the Unique Structure Forming Aesthetic Philosophy, Around It's Creation, To Give Intellectual Context - Recollections, Which Are Painful To Discuss, Yet More Intolerable, Undisclosed - Thus Testifying, in My Creator's Capacity, to Give Creedence, To The Innovative Design System, That Was Hitherto, "Lost In Translation" - My Conscience Has Been Cleared, As I Wrestled Considerably, In The Writing Process - To Explain the Technical Subtleties, Which Were Expurgated and Trampled Upon - An Isotonically Dynamic, Architectural Engineering System, That Neutralizes, the Deleterious Effects, of Load Baring Gravity - In The Spirit of Rational Discourse, I Have Carefully Went, About My Business - At Once, Offering Critical Analysis, To Purloined and Mystified Things

Justified At Times, Along The Way, To Vent My Spleen, Upon Those Who Wronged Me
Their Psychology, In Doing So, Unfathomable To Me - Uncreative Human Beings,
Are Often Driven, To Destroy Many Beautiful Things - Artistic Frustrates, Have Been
The Bane Of The World, From Beginning of Artistic Expression - They Murder To
Create, Their Grand Palette, of Annihilation, The Wasted Blood Sacrifice, of Unjust War
Entangled In Web Of Lies, Fearing Encroachment, of Lurking Spider - Who Feels
Their Futile Thrashing, Upon Chaotic Web - The Spider Alone, Knows How To Navigate,
The Insect Catching Trap, Having Created it, in Instinctive Secrecy - Patiently Waiting,
For The Victim, To Become Caught Up Within It - Primordial Ancestors Unchanged,
Caught, in Hardened Tree Amber, Testify to the Enduring Brilliance, of This Ultimate
Survival Strategy - The Spun Twining, Of The Three Sister Fates, of Greek Mythology:
Correspond Directly, With Corresponding Reality - One Weaves, One Twines, Another
Cuts The Strings - With The Cold Precision, of Black Widow Spiders, They Set Snares,
And Boundaries, To Curtail Human Folly, and Overreaching Hubris - Preventing Chaos
And Anarchy, From Taking Hold, in the Ordered Progression, of the Universe - Only
Those Who Know Their Ultimate Path, And Remain True, Without Compulsive Trespass
Achieve Safe Passage, Through The Dramatic Pathos, Of Tragical Entanglements -
Some are Made Example Of, Allowed To Raise Up, to Unhallowed Heights, Made To
Suffer Indignation, Of Greatest Fall, From Grace - So That Those Who Witness,
Such Sardonic Spectacle, Give Second Thought, To Impulsive Acts, of Unreflective
Wickedness - Some Made Blind, By Hoarding Greed, They Are The First To Fall:
Magnificent Conglomeration of Seven Deadly Sins and Dust, Splayed Upon the Ground

The Alchemy of Transmutation, Can Turn One Precious Thing Into Another - Turn
Plebum Lead, Into Alchemic Gold - Bring Down The Power of Heaven, To Complete
The Great Work - From Such a Place, All Things May Come - Dreams Manifest, Into
Bold Reality, Graced By Inspired Imagination, Conceptualize in Prophetic Vision, And
See It Through - Sacred Rite, of Inventive Process, Bestowed Upon Genius, With Open
Mind - New Knowledge Springs Forth, From Such a Source - To The Daring Few, Who
Would Bare It, Upon Their Straining Shoulders - Nurtured Brain Child, Into Existence,
With All The Caring, of Proudest Parent - To See It Mature, into Powerful Conception,
Until Readied To Leave, A Thing, Unto Itself, Self Possessed, Becoming of The World
Such Ideas Remain, Or Return To Dust, They Come When Needed, To Help Us -
Unbroken Chains of Progressive Innovation, Define The Health of Evolving Culture
In Contrast, Closed and Autocratic Societies Implode, Censured Creativity, Stifled -
Plagiarism Speaks Directly To This, The Modus Operandi, Piracy of Free Expression
Indicates, Intellectual Vacuum, Cultural Desert, Intellectual Purgation, Criminal Trespass
A House Thus Bankrupt of Innovation, Will Not Stand, Those Who Create, Driven In
Exile, From The Blighted Land - Their Prognosticating Warnings, Ignorantly Unheeded
I Have Seen Such Ancient Archeology, Long Abandoned, Modern Ancestors
Grown Primitive, Aboriginal, Sell Trinkets, Before Their Forefathers Glorious Monuments
The Ironic Contrast, Was Not Unobserved - I Thought of the Devolving Process, Which
Led Such Things To Transpire - Always There Was a Purging, of the Best and the
Brightest, Religious Orthodoxy and Caste System Ossifying, Halting Vertical Social

Movement - Food and Water Shortages, Bring Malthusian, Population Culling - Free Thought Philosopher, is Replaced With Propagandist and Demagogue - Thinking Discouraged, Producing Dissociative, Dream Reality - Then One Day, The Means of Production, for Civilization Maintenance, Simply Disappears - No One Left To Maintain, The Infrastructural Machinery - Ruins of Great Cities, Become Shepherds Pastures - Masterpieces Sold, as Worthless Stone, To Archeologists, By Illiterate Peasants - What Took A Thousand Years To Build, Often Leveled in a Single Generation, In Perfect Storm

I Shall Not Be, The First Nor The Last, To Have My Creative Manufacture, Abused and Stolen - Perhaps Remembered, For Being Most Vociferous, and Philosophically Focused, Upon The Matter of Plagiarism - Beyond the Act Itself, The Symptom of the Age and Declining Culture, It Thrives In, What Is Says Of, the Role of the Artist, Becoming Disposable, Made Invisible - The Diminution of Creative Process, To a Meaningless Element of Production - Placing the Mercantile, and the Constructed Object, Before the Conceptualists, Original Idea - The Ramifications in Perspective, are Telling, The Role of the Autonomous Individual, in Corporate Society, in Jeopardized - An Unknown God, a Juggernaut, That Knows Not Our Name - Such Sanitized Society, Bereft of Human Connection, The Objectification of Everything, That Would Be Art - Isolates and Insulates Us, From Each Other and Ourselves - Soon Unrecognizable, As We Gaze Into The Mirror, Of Our Age - With Secret Leadership, Hermetic Cabals, Organized Criminals, With License To Steal - Where Are The Geniuses of the Culture Where The Age, We Can Reference, Study and Emulate - Like the Middle Age of Europe, A Thousand Year Darkness, Anonymous Art, Adorns Countless Cathedrals, Nothing Advancing, No Schools of Free Thought, Servile Organs, To Totalitarian Church And State - With Freemason Sculptor, The Only Traveling Master Craftsmen, Yet Still Anonymous - There Is Such Danger, When You Kill The Identity, of the Creative Artist - Dark Ages of Arrested Culture, Must Be Beaten Back With Illumination - They Are Not Accidents of History - They Come, When Arts and Sciences, Have Been Exploited, Suppressed, Ignored, Neglected, and Finally Die - It Happens At Any Time and at Any Place - Where Fundamentalism, Orthodoxy, Chauvinism, Ignorance, Abuse of Power Take Hold - First Against, The Free Thinker, The Intellectual, The Creative Artist, The Inventor, The Scientist - THEN EVERYONE - Under Pretense of Security It Enslaves, By Censuring Culture, It Blasphemes, Through Suppressing Philosophical Discourse, Collective Consciousness, Becomes Deranged - Intellectual Theft, Violates The Public Trust: Accelerating Epidemic Ignorance, Where Original Genius Vision, is Thus Obscured, Under a Mountain of Counterfeits, It Deludes Itself - Untouched By Such Authenticity, a Synthetic Reality Clouds, Aesthetic Perception - Empathic Connection, Remembrance Of Things Past, Dissociates, Made Mockery, In Pirated Imitation - Soon Oblivious, We Ourselves, Becomes Assimilated Victims, of Such Consumption, Become Synthetic Approximation, Of What We Would Be, Self Actualized - What Aesthetic Guides, To Calibrate and Adjust, Our Critical Vision - Long Copied Out, From Stolen Archeology, Replaced By Pirated Fakeries, Generations Before Us - I have Thus Shared, My Unusual Story, My Unique Perspective, For I Have Seen This Coming, Close At Hand - Philosophers Often Anticipate Trends, They Have First Born Witness To - Before the General Public, Comes To Know Such Things - Some Say

They Are Even Born, Like Their Brother Epic Poets, From Out, Such Tumultuous Times, As Form, Of Circumspection Remedy - Offering Method To Repent, Severed Head Served Upon a Platter - Still Others in Horangued Exile, Must Sing For Their Supper - Dodging Assassins Blade, of Grazing Bullet, Made Strangers in Their Own Land, Unrecognized, Though Native Born - Becoming Citizens, Of The World, Wherever Truth Triumphs, Over Bombastic Rhetoric - Few, Are The Philosophers Welcome, In Their Own Place and Time, Likewise The Eternal Bards, Who Define Their Culture, in Epic Poetry - To Harsh, a Light They Cast, Upon Corrupting Depravity, Their Reality Too Shocking, For the Grotesquery Culture, They Would Reflect Upon, And Thereby Cure - Martyrized, a Murdered Reliquary, Silence Golden, During Bonfire Book Burning - Soon The Great Libraries Follow, Next The Houses of the Holy - Chain Reaction Of Events, The Tyranny, of The Ignorant Populace, Shows No Quarter, Nor Mercy Given Immersing Themselves Further, Into The Quagmire, Of Dark Centuries - All Beginning With the Single Trespass Infraction, Against the Sovereignty, of Individual Intellectual Property - Soon The Plagiarist, Must Play The Impostor, The Impostor Become Blundering Dilettante, Multiplying His Crimes, In Egotistical Self Defense - Deluding Others, as He Perpetrates, Fraudulence and Sham, He Must Speak Upon, His Stolen Masterpiece, With Contrived Authority, To Authenticate His Lie - False Teachings Rendered, Like False Prophesies, Have a Tidal Cultural Effect - Reality Itself, Is Cult Undermined, Understanding And Knowledge, Made Bankrupt, Proof Unverified - Like a Missing Rung, Upon a Ladder, A Tool of Ascent, Denied - The Evolution, of Progressive Thought, Is Interrupted And Halted - Quantum Leap, Is Then Required, Approximation Of Acrobatics, Perilous - Enough Rungs Go Missing, The Ladder Will Not Support, Nor Stand - No Direction To Go, But Downward - Cascading Backward, To Stone Age - ALL Starting With Bold Acts of Plagiarism: Original Genius Left Uncredited, Mystification, of Scientific, Massaged Fact, Becoming Pseudo-Religious, Fiction, Orthodoxy Crushing, Open Discourse, Questioning Inquiry, Stifled - Jangled Propaganda, Warping, Deep Fathomed Philosophy - Disingenuous Professionals, Become Pawn Brokers, of Stolen Articles, And Embellished Wares - Left Destitute, Infrastructure Gone, Devolving Back, To The Ancient Ways, Of Primordial Ancestors, Civilization Gone - The Final Purgatory, Of Stolen Ideas, Metastasizes, Until We Are Left With Nothing, But Legend of The Fall Remember, The Greatest Manifesto, Of The Magician, Magnus Adeptus, and the Master Polymath Alchemist, Transmutationist - "I Am an Icon, Because, I Can Do Anything" - Nothing, Beyond Their Capability, Given Grace, Will and Time: The Original Imagination To See Beyond What Has Been - To Capacity, Through Increased Technology, To Bring It Forth Into Being - These Miraculous and Magical, Deeds, Acts, Afforded To The Few Trespass Against, This Sacred Ability - You Are Left, With Mere Fragments of Plagiarist, Faked Artifact, Fabricated Archeology, Stolen Scientific Theory, Infringed Patent, Counterfeited Manuscript, Copied Masterpiece, Assumed Identity, The List is Endless, The Mentality The Same - To Take What Is Original, Great, Energetic, New, Unique and Revolutionary - Rendering It Useless, By Senseless Copyist - Depriving All Involved, With The Original Work, Substituting It, With Inside Joke, and Pretenders Farce - Is There Enjoyment In Perpetrating a FAKE - I Suppose Only, If The Fakery Secret, Is Maintained - Until Then, It Is Highly Hysterical: "Like Two Dirt Farmers, Fighting Over

Dirt" - An Ironic Quotation, From My Former Lawyer, Robert Bergen, Now That The Shit Is Hitting The Fan - I Wonder What Other Choice Words of Wisdom, He Has Gathered For My Edification - Perhaps He Can Write a Book, Failed Attempt, To Explain His Dirt Farmer, Lapse of Reason: While Conflicted of Interest, Functioning, as Corporate Attorney of Record, Pimping Out, My Designs, To Silverstein, MTA and Port Authority To Those Who Found My Skyscraper Design, To Your Aesthetic Taste and Exceeded Expectation, Spending Five Billion Dollars, To Build Your Ivory Tower - Never Considering Design Fees, or Engineering Renumeration, Of Any Kind - Consider This Published Book, Delayed Invoice, With Compounded Interest, For Your Criminal Negligence, and Intellectual Theft: Hereby, You Are Officially, Put On Public Notice, As Plagiarists - I Am Sure, The Trans Hub Design "Appropriation", Has Nothing To Do, With The MTA, VP of Legal Affairs Job, Bergen, Currently Occupies, as: "Resident Empty Suit" - Enjoy It, While It Lasts - What Goes Up, Must Come Down - Reputations and Careers, Put Into Jeopardy, From Appropriated Design Work, Solicited and Plagiarized A Pound Of Flesh Stolen From My Heart, A Bison Rheum, Upon Your Beard -

Honoring The Ancestors, In Their Triumphs and Their Glories, Honoring The Creator That Thus Created Us - Paying Tribute To Their Honored Memories, With Righteous Reverence, The Genetic Line Unbroken, Through Cellular Memory: Distended Line, Of Philosophical Reasoning, Metaphysical Discourse, Often Heard In Whispers, in Prophetic Dreams - This Is Where The Origin, of Imagination Arises, This Is Where, Long Pondered Thoughts, Come Forth Into Reality - Creativity, Thus Derived, Is The Labor of Generations, Not Limited To The Confines, Of a Single Life, Or Individual - Past On, In Living Legacy, From Where We Come From, Determining in the Final Analysis, Where We Shall Be - Evolutionary Progress, The Course of Things, Tethered To This Conception, of Intellectual Immortality - Thoughts, That Are So Aligned, With Reality, They Will Not Die - Spanning Across The Universe, The Mind Wanders, Ceaselessly - Questioning, The Nature Of Existence, That We Are Bound By -
The Provenance of Knowledge, Our Human Heritage - Unique, Unto Ourselves, To Ponder, The Mysterious Meaning, Of Our Lives - No Other, Must Therefore Trespass, On What, The Muses Endow, What The Gods Inspire, What The World, Would Have Us Understand - To Circumvent This Process, Is Anathema, Obstructing It's Own Divine Navigation, Is The Primrose Path To Hell - Well Intentioned Philosophies, That Espouse, The Good, Of The Many: The Tyranny of Crowd - Overlook This Archetypal Convention, From Where Liberating Dreams, May Come, Overlooking The Individuated Process, Of The Creative Being - They Bring, To Gradual Ruination, Formation of Civilization, Murdering The Messenger, Killing The Golden Goose - Argument Is Useless, At The, Tipping Point, Of Decline - Words Lost, Upon Nihilist Sychophants, That Eviscerate Innovative Things - Past the Point Of No Return, They Horde and Conserve, What Is Irrelevant and Arcane: Leaving Breadcrumbs, For The Masses, Like Chattel, To Barely Sustain - Nature Abhors, A Vacuum, History Rebukes, The Ignorant and Weak, Evolution Favors The Strongest Mind, Through Natural Selection - The Epidemiology, Of Plagiaristic Blight, Cult of Personality, Dissimulation and Impostory, Unchecked and Undetected, The Chronic Intellectual Disease, Of Culture In Decline - No Oracles Speak, No Auguries Portend: Left In Existential Psychology, The Unimaginative Dwindle

What Brought Us To This Crossroad, Will Be The Reflection of Historians, Archeologists Will Shudder, As They Calculate, The Rate of Fall - Baffled By The Way, In Which The Creative Purgation, Took Place - Superseded By Egotistical Vanity, Of Dominating Oligarchic Few - The Paradigm Shift, In Consciousness, Too Late And Among Too Few, When The Dam Broke, And Niagara Burst, No One Left, To Reach Higher Ground: Holy Mount Ascent Forbidden: When Wicked Men Arose, Releasing Dogs of War, Murdering Messenger, Silencing Escaping Prophet, Passing The Border Guard, To Barbarian Periphery - Leaving Farewell Message, In Digested Treatise, No Longer Able To Elaborate, In Censured Forum - Exasperated At Speech Curtailed, Upon Water Bison, The Ancient Sage, Sets Upon The Journey, To Barbaric Lands - Rather Than Disgrace The Effete Sensibilities, Of The Doomed Over-Civilized - They Who Build Walls, Limit Travel, Contain Movement, With Arrogance and Swagger - Freedom Reserved, Behind Closed Doors, For Certain Few - When The Fires Burn, and the Riots Begin, Oblivious, They Entertain and Amuse Themselves, In Fortified Towers - The Unwashed Masses, Perceived, As Ants Upon The Ground - Their Agony Ignored, From Such Altitude - Like Babylonian Spire, The Avenging Lightning Will Strike, Tower Will Tumble, The Arrogant Fall - The Masquerade Party Ends, Identities Revealed - The Guillotine Tables Reverse, The Process Begins Again - Ridiculous Cycles, of Rise and Fall, When The Great Work, Lies Fallow, and the Vain Egotists, Flourish - Separate Narcissistic Reality, Spurning And Condescending, To The Creative Nature, That Moves The World: Evolved Vestigial Beyond The Point of Breeding, Inventing, Creating, Independent Thinking - They Simply Consume, Like Vampiric Parasites - Like King Gilgamesh, of UR, Observing The Floating Suicides, From High Rampart Walls - Longing For ENKIDU, His Wild Man Half, Prophesied Boon Companion, To Bring Him Out, The First Civilized, Man Made Hell - Like Autumnal Leaves, Like Pandemonium Legions Falling, Into Raging Rivers, Live Volcanic Pools - Earliest Ancestors, Felt Imprisoned, In Cosmopolitan Dwelling - Balance Must Be Made, Between Decadent Civilization, and Anarchic Barbarity Clearly, There Is Creative Room, For Many Ways of Being - Regimentation to Anything, Beside Ones Philosophy, Is Patently Absurd - Once Again, the Pirating Plagiarist, Limits Our Potential Insights - The BEST Creative Inspiration, Comes From The Methodical Academic Study, of ORIGINAL Work - Such a Premium Placed, Such a Golden Standard Established, It Raises The Aesthetic Bar, And Creates Generations of Masters By Contrast the Stolen Work, Exudes the Ignorance, of the Meaningless, Has No Resonant Credibility, Reveals No Higher Truths, Pays No Culture Tribute, Concealing Confabulation of Lies - An Insult To Intelligence, It Smacks of Artificiality - Even Compels The Plagiarist, To Hide Their Face - Longed For Dubious Credit, Becomes a Reminding Curse, The Bankruptcy of Aesthetic Ideal, Professional Ethical Degradation - I Pity Those, Laid Those Low, To **"Thought Cannibalize"**, The Original Work of Another: Like Richard Plantagenet, In His, Winter Of Discontent, Descanting Upon His Own, Hunchbacked Deformity, in the Shadow of the Sun, So That Dogs Bark at Him, a Freak Of Nature, While Made Crooked, He Perambulates, Meditating, Within Twisted Mind - To Self Overthrow, The Royal Dynasty, Last Of His Line, Botched Spectacle, Counterfeit Specie, Of Once Great Warrior Line - Preferring the Intrigue, of Assassination, Against Obstacles, To Throne Ascent - Killing Imprisoned Brother, Burying His Young Wards, Under Steps, Of Tower of London: All Machinations, To Mar The Throne of Scone, Pollute Uniting Scepter, of **WASP** England: Black Aristocracy, "Raiders" Soon Follow

# REALITY A WORLD OF VIEWS

## The Strange (and Sad) Life of Bob Diamond
by Simi Horwitz · February 8, 2015

"In person he listens carefully to Diamond's comments, interrupting him with a gentle, reassuring pat on the arm, "Bob, Bob, let me," when he feels Diamond is getting agitated or not responding as clearly as he should to my questions.
Castillo and Diamond have been pals since their days at Midwood High School. "Bob was nerdy and nervous, but some kind of genius," recalls Castillo. "In 12th grade Bob won an alternative energy sources competition for designing a solar cell satellite that converts solar cells into electricity, which could [in theory] run three cities." Castillo says that he too was a science wiz (in addition to being an epic-poet and sculptor) and both men were **"informally involved, though at the highest civilian level, with the FBI and/or NSA [National Security Agency]."** "Before you think Greg and I are crazy and having shared delusions–" Diamond mutters as Castillo finishes Diamond's sentence "–**It's not unusual for high achieving math and science students to be recruited. It's intellectual feudalism.**"

**DEPARTMENT OF THE AIR FORCE**
OFFICE OF THE CHIEF OF STAFF
UNITED STATES AIR FORCE
WASHINGTON DC 20330

APR 12 2010

HQ USAF/CC
1670 Air Force Pentagon
Washington, DC 20330-1670

Mr. Chester Burger
33 West 67th Street
New York, NY 10023-6224

Dear Mr. Burger Chet

On behalf of the men and women of the United States Air Force, please accept my thanks and gratitude for your years of volunteer service as a member of our Public Affairs Advisory Group in New York City. For 15 years, your highly valued guidance helped advance our national security by strengthening public support for vital Air Force programs, strategies, and missions.

Specifically, your informal leadership of the advisory group, to include recruitment of some of New York City's brightest public relations minds, brought unparalleled advice and counsel directly to Air Force Secretaries Sheila Widnall and James Roche, five chiefs of staff, and numerous flag officers and public affairs leaders. Through your assistance, we have established lasting relationships with leading New York City opinion leaders and organizations and national print and broadcast news agencies – relationships which continue to foster influential support for Air Force and Defense priorities.

I salute you as a lead figure in shaping the global broadcast news and public relations professions, as a volunteer advisor to our national intelligence agencies, and as a former World War II Airman. The channeling of these experiences – of your extraordinary communication knowledge and skills – significantly improved our Air Force and contributed to our Nation's defense. We thank you for your selfless, patriotic service to your country and the United States Air Force.

Sincerely

NORTON A. SCHWARTZ
General, USAF
Chief of Staff

# FORM VA
For a Work of the Visual Arts
UNITED STATES COPYRIGHT OFFICE

127830072

## 1 TITLE OF THIS WORK
NEW WORLD TRADE CENTER STRUCTURE  ARCHITECTURAL WORK

## 2 NAME OF AUTHOR
a GREGORY CHARLES CASTILLO

DATES OF BIRTH AND DEATH
Year Born: 1960

Author's Nationality or Domicile: UNITED STATES
Domiciled in: UNITED STATES

NATURE OF AUTHORSHIP:
☑ Architectural work

## 3
a Year in Which Creation of This Work Was Completed: 2002

## 4 COPYRIGHT CLAIMANT(S)
GREGORY CHARLES PHILIP CASTILLO
234 VAN BRUNT (1st FLOOR)
BROOKLYN, NEW YORK 11231

APPLICATION RECEIVED: JUN 4 2002
ONE DEPOSIT RECEIVED: JUN 4 2002

Delivery Receipt

From: BHRA
499 Van Brunt Street, #3A
Brooklyn, NY 11231

To: — 1 tube (w/ WTC Design Plans inside)
to be delivered to:

Mayor Bloomberg

— 1 tube (w/ WTC Design Plans inside)
to be delivered to:

Deputy Mayor Doctoroff

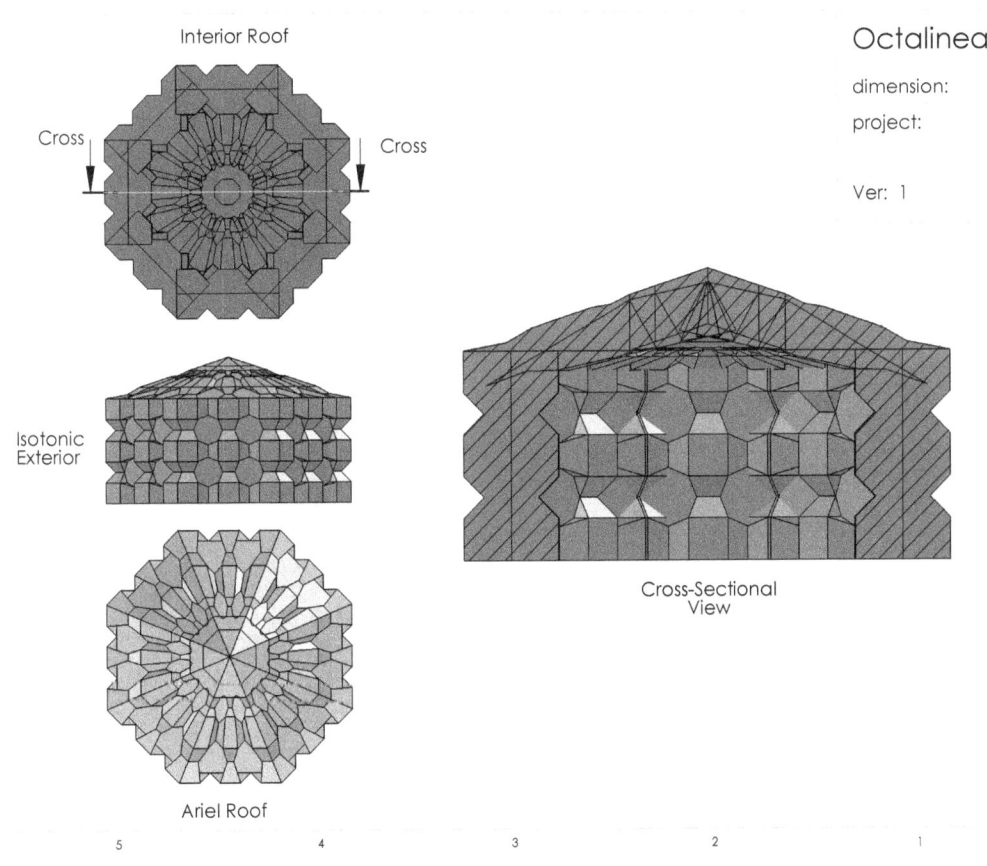

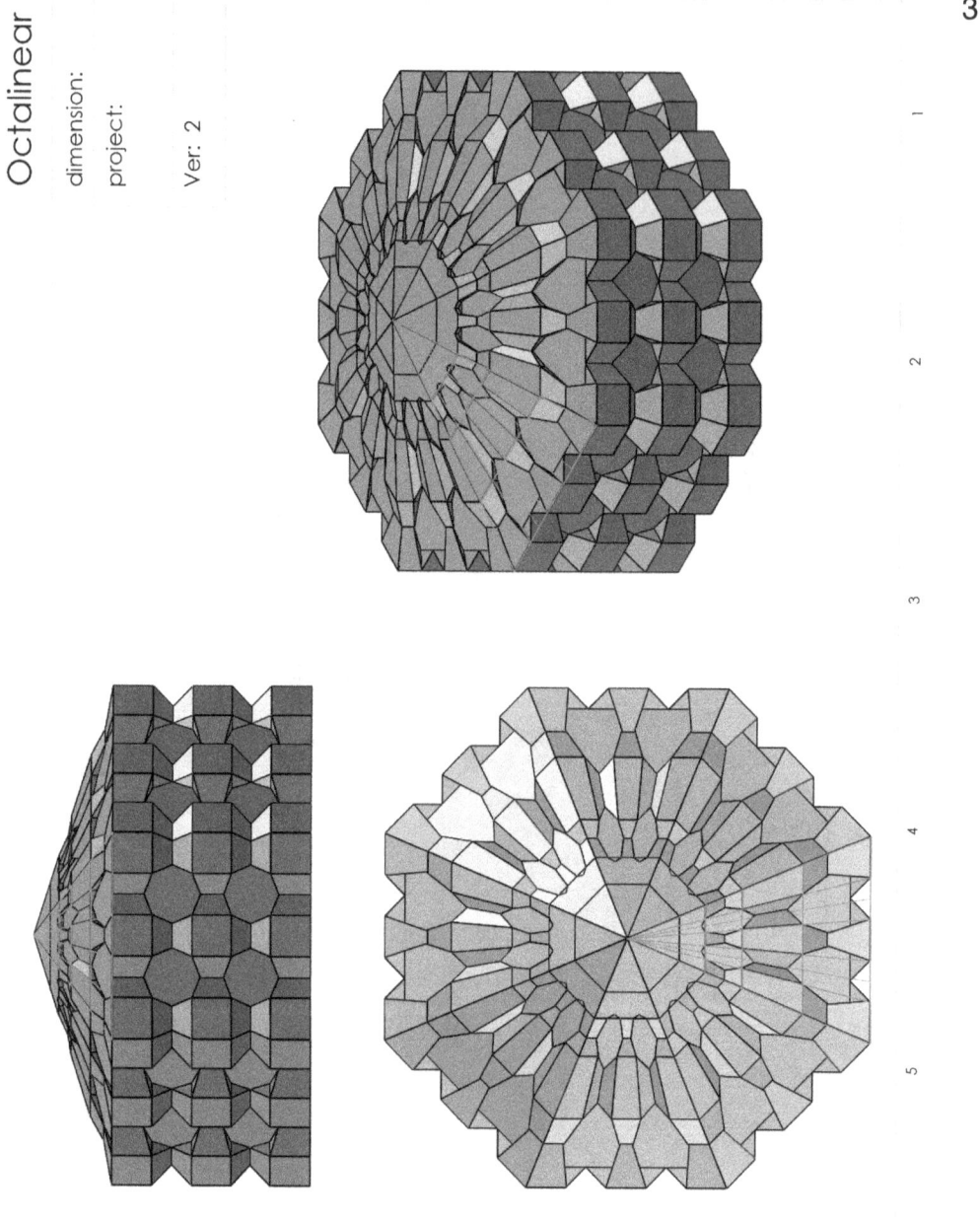

# Octalinear

dimension:
project:

Ver: 1

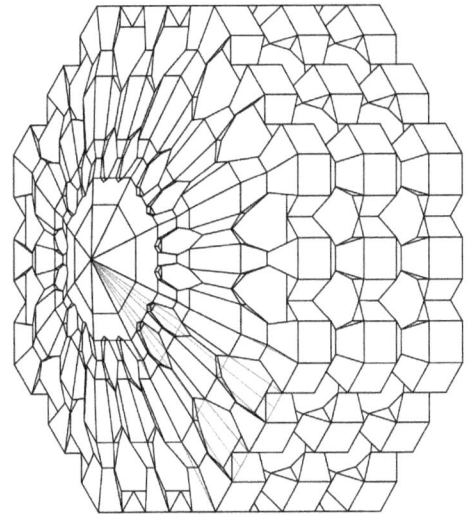

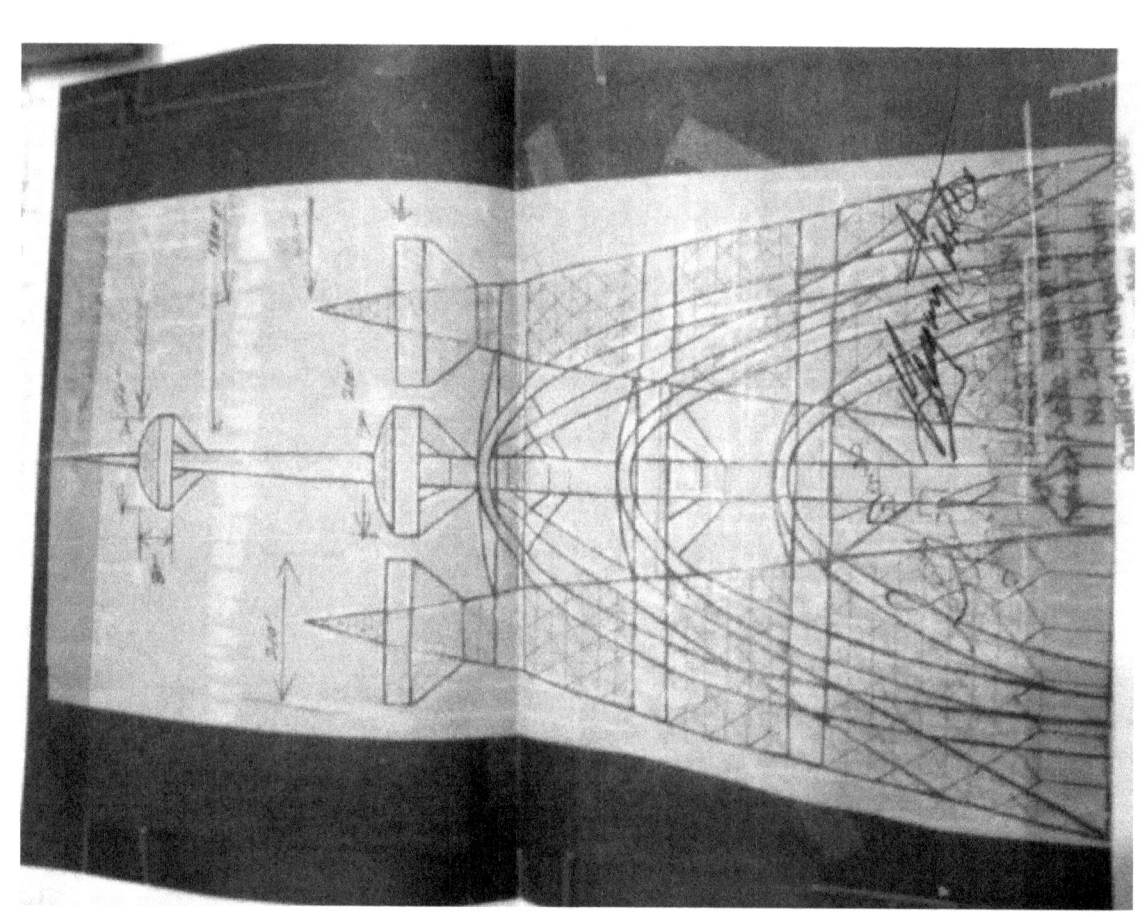

Photo # NH 58769  Cutaway drawings of the Confederate submarine H.L. Hunley

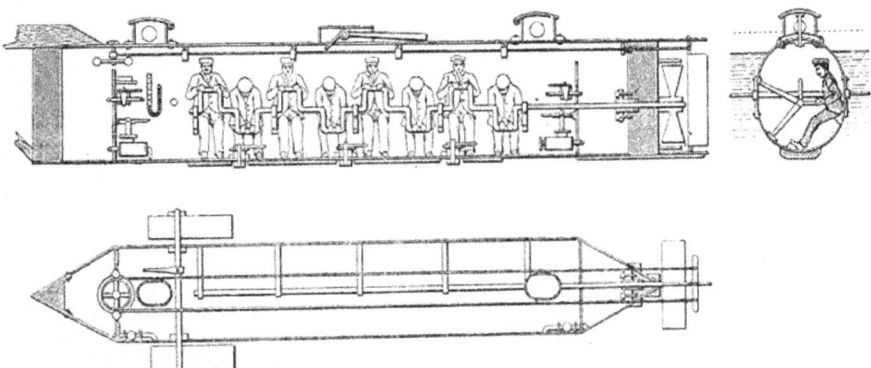

Fig. 175 à 177. — Le *David* de Hunley reconstitué d'après les dessins de M. William-A. Alexander (1863).

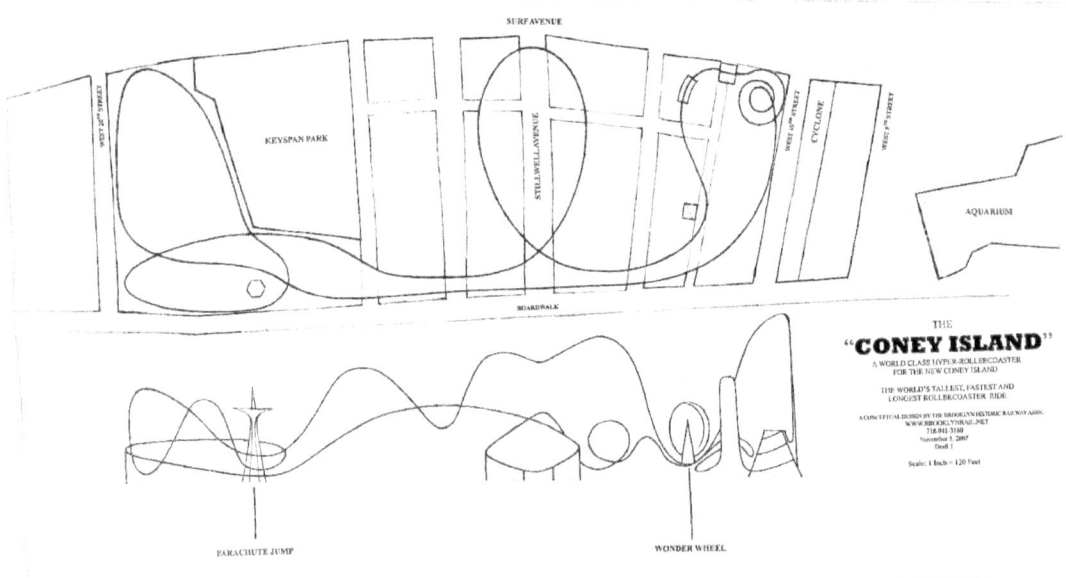

# TO WHOM, IT MAY CONCERN:

It has recently come to my organizations attention, after witnessing the new construction of World Trade Center, "Building One" - That in fact it bore a striking resemblance, to a plan we submitted to the Lower Manhattan Development Corporation (LMDC) an incorporated division of NYC EDC, approximately ten years ago, as the Design Initiative Group (DIG) the design wing, of Brooklyn Historic Railway Association (BHRA) The plan in question was originally conceived and developed by myself, Gregory Castillo, President, CEO of BHRA, Designer, Robert Diamond, Chairmen, BHRA, Project Manager, over a nearly six month period, requiring hundreds of our dedicated man hours to develop, research and technically accomplish.

The request to initiate said project, was in the form, of a verbal contract, between Peter Brightbill, esq. and myself, Gregory Castillo, of BHRA, in the legal offices of Robert Bergen, esq., a Partner of Holland and Knight, Peter Brightbill and Robert Bergen, were the co-counsels and corporate attorneys of record, for BHRA at that time. The verbal contract for submitting a design to the LMDC, was also witnessed by Robert Diamond, of BHRA, Robert Bergen, H & K, Frederic Papert, of The New York Landmarks Conservancy, who were all in clear earshot, and privy to the entire conversation/verbal contract, entered into between, Brightbill and Castillo, at the legal offices of H & K.

Unbeknownst to BHRA/DIG executive staff, was that Peter Brightbill, esq., was also a legal consultant/contractor for the LMDC, and as such solicited our design skills, sweat equity and good will, with a clear and undisclosed legal conflict of interest. The submitted BHRA/DIG plans and studies, were submitted by Peter Brightbill, to LMDC for their consideration, and also sent directly to the Mayors Office by BHRA/DIG, via a designated BHRA Staff Member, who requested and received, a time stamped receipt from same.

The BHRA/DIG WTC plan, then rematerialized, in the major press, in a modified form, as plan number three of Skidmore, Owens & Merrill (SOM), and David Childs, Architect of Record, initially proposed design sets, submitted for the New World Trade Center. My organization was never mentioned or given any design credit whatsoever, nor compensated for same, in the advertised SOM initial design, or selection process, or at anytime thereafter.

The BHRA/DIG, completed the unique and original, dual (twin) pyramidical, obelisk design, incorporating elements of the original twin towers architecture (ground level), inventing an innovative steel system understructure, provided and fully explained, in numerous email explanatory notes and texts: establishing clear design provenance, with detailed architectural renderings, and professional architectural illustrations, provided in multiple emails and technical attachments - Sent around to numerous New York City officials, for comment and design/engineering feedback, over the entire course of the six month design process, all of which was fully archived and documented, by BHRA executive staff.

Furthermore, BHRA/DIG providing a full and complete design program, in a good will attempt, to help get the New World Trade Center, an important national symbol, off the drawing board and onto the New York skyline - Now the partially compromised SOM/David Childs architectural design, reduced from the BHRA/DIG, surreptitiously appropriated, original "dual pyramid concept", subsequently analyzed and extrapolated, by Cantor Sein.UK, into the current engineering plan, and new design and construction concept, for WTC "Building One", constitutes a clear Intellectual Property and Copyright Infringement issue, against my organization, BHRA/DIG.

We have also recently discovered that WSP CSUK, has recycled the appropriated BHRA/DIG skyscraper building design, once again, this time, in the newly constructed "SHARD" building, in London, England - In which they have taken design/engineering credit. WSP CSUK, demonstrating a clear pattern of total disregard, for the intellectual property rights, of BHRA and it's ad hoc design wing, DIG/Design Initiative Group.

My organization was informed in 2002, that the BHRA/DIG WTC design, was selected to be wind tested, at the University of Chicago, and that an exact scale model was subsequently commissioned, to study the structural integrity of the BHRA/DIG design program, in simulated hurricane force winds - We were later informed that our unique twin skyscraper design, had indeed passed with flying colors, exceeding all speculated engineering expectations. That is in fact, the last time my company ever heard, of our original design being considered, for potential construction, through any reliable primary sources, and email established official channels, developed during the design phase of the BHRA/DIG WTC project.

At some point during the summer 2004, BHRA, President, CEO, Gregory Castillo, met an employee of WSP Cantor Sein.UK, at a Pratt Institute alumnus gathering, the WSP CSUK employee, in the course of our conversation, indicated that he was an Architectural Engineer, then working on the New World Trade Center design - Upon further inquiry, Gregory Castillo, determined that the design that WSP Cantor Sein.UK, was actually working off of, to analyze the structural steel system, that it was contracted and commissioned to engineer, for the new WTC skyscraper, was in fact the BHRA/DIG original "twin pyramidical design". The employee of WSP CSUK confirmed my own correct suspicions, saying that he had actually seen, the "DIG" graphics, at the lower left hand marginalia, of the architectural illustration sets, that they were then studying, at the architectural engineering offices, of WSP CSUK.

Castillo, met the same WSP CSUK employee, at Pratt Alumnus gathering the following year, and the architectural engineer once again, confirmed that WSP Cantor Sein.UK, was still working on the planning phase, of the unique steel understructure, attempting to calculate the steel schedule required, to structurally support a limestone facia edifice, attempting to match the Empire State Building, exterior masonry treatment; A structural engineering plan, they apparently later abandoned, due to load baring weight concerns, discovered through extensive computer simulation analysis, a technical fact, corroborated by the same WSP CSUK employee, to Mister Castillo, at their second meeting - At the subsequent Pratt Alumnus gathering, in following summer of 2005.

BHRA/DIG was also later actively solicited, by the Brooklyn Borough President's Office, Marty Markowitz, requesting our organization to submit plans for the New Coney

Island Design, in 2007-2008, by his official letterhead, submitted as an endorsement and overture to the local Coney Island LDC and NYC EDC officials, for our group to provide original design plans, for a "Super Coaster" as well as professional research, for soliciting top international Amusement Ride manufactures and vendors, again without any recognition or renumeration of any kind, for planning and research services rendered.

BHRA/DIG also developed, a full intermodal transportation plan, for Coney Island, extensive land use suggestions, for year around tourist attractions and water park themed hotels, revamped commercial strip boardwalk, ocean pier and beach front re-development and revitalization, through re-establishing water ferry historical routes, all fully documented and archived, by BHRA executive staff.

BHRA/DIG provided recommendations for professional architectural illustrators, from the Sandborn Map Company, for the complete "Orthoimagery Projection" architectural renderings of the BHRA/DIG Coney Island Master Plan, which was later published in the major media - Once again, our design group was not given design credit, or remunerated in any way, shape or form - Yet another five month Herculean design and research effort on our part, fully documented by BHRA/DIG, in extensive email traffic and attached architectural illustration and design email strings, to the CILDC and NYC EDC Officials, over the entire four month Coney Island design process, in question.

My organization has requested me, as president and CEO, of BHRA/DIG, to contact you on this serious matter. Because of the extraordinary length of time that has transpired, since the verbal and written contracts, were entered into, and the BHRA/DIG design programs were executed - And now the recent commencement, groundbreaking, and actual building process, of the new architectural constructions, in question, with there "Striking Similarity" to our original and unique concepts and designs, the established, documented and clear chain of custody of the original BHRA/DIG designs, by our former attorneys, and subsequent reliable anecdotal reports, of them being fraudulently vetted and utilized, by the several contracted, architectural and structural engineering firms aforementioned, WITHOUT, BHRA/DIG's, expressed written or verbal permission, to ever do so.

My organization is very well aware of the embarrassing subtleties, political delicacies and professional status, and ethics involved, the potential ramifications of this important legal matter going public, therefore, in a good faith effort, we are reaching out to you, and are simply seeking out, our well deserved design credit, and an appropriate monetary renumeration, based upon our significant contributions, on these three major design projects.

Of course with this first overture and equitable gesture, to resolve these complicated issues in question, and further bearing these in mind, the reasonable considerations, of our fair requests for monetary renumeration and entitled design credit: BHRA/DIG executive staff, are also factoring in, the overall cost and technical complexity, of their original design program efforts, and intense involvement, with regard to the a overall planning and research process, in general. Significant contributions, that in fact made the current WTC "Building One", "SHARD" Building, and New Coney Island original designs, all aesthetically and technically possible, in an enormous concerted organizational effort on our part, that occupied almost a year of BHRA/DIG dedicated design and uncompensated and unacknowledged planning time, without being given any credit or monetary renumeration whatsoever, before, during or after the development, of these three major, high profile, world class design projects.

    Our organization, BHRA/DIG, and I, look forward to resolving this issue, as soon as possible. We look forward to arriving at an amicable and fair design credit and monetary resolution, for services rendered, to this impending Intellectual Property, Copyright Infringement, and fraudulent unfair use, of BHRA/DIG designs, problem. Please have your authorized corporate principals, or retained legal representation, contact us through our company website www.brooklynrail.net , at your earliest possible convenience, so that we may commence earnest negotiation efforts, to rectify this outstanding, and potentially legal, and most unfortunate situation at hand.

Sincerely Yours,
Gregory Castillo,
President, CEO, BHRA/DIG
cc:
Skidmore, Owings &Merrill, David Childs, Architect, SOM
WSP Cantor Sein. UK
Sanborn Map Company
University of Chicago, Department of Architecture
President, Economic Development Corporation (NYC)
President, Lower Manhattan Development Corporation
President, Coney Island Development Corporation
President, Port Authority of New York and New Jersey
Peter Brightbill, esq, Corporate Counsel, BHRA, Tishman Construction, LMDC
Robert Bergen, esq. MTA Executive Vice President (Formerly of Holland and Knight)
Marty Markowitz, Brooklyn Borough President
Andrew Cuomo, Governor of New York State
Mayor's Office, City of New York, Michael Bloomberg
Robert Steele, Deputy Mayor, City of New York
Hardy Adasko, EDC, City of New York
Silverstein Realty Group, WTC
Tishmen Developers, Building 7, WTC
Frederic Papert, The New York Landmarks Conservancy,
42nd Street Development Corporation
Robert Diamond, Chairmen, Brooklyn Historic Railway Association:
Project Manager: BHRA/DIG
Gregory Castillo, President, CEO, Design and Construction, Planning, BHRA
Brian Kassel, Vice President, Urban Planner, Design Assistant, BHRA
FBI, "Public Corruption Unit", New York City, HQ, Federal Plaza
Gabriel Salem, esq, BHRA Corporate Attorney (Present)
Jay B. Itkowitz, esq. BHRA Corporate Attorney (2002-2003)
Lower Manhattan Development Corporation, (LMDC) NYEDC
Santiago Calatrava, Architect, LLC, MTA WTC, "Trans Hub"
Renzo Piano, Architect, Scethlia Shard Building, London UK (CSUK)
Alexander D Garvin, Architectural Professor, Yale School, of Architecture, LMDC

# SACRED GEOMETRY:

Sacred Geometrical Architecture, The Second Part, of My Philosophical Thesis - To Balance Out, the Unfortunate Subject, of Architectural Plagiarism - With a Philosophical Discourse, Closer to My Own, Temperament, and Avocational Passion, For Architectural Design and Industrial Engineering - Generally I Have Restricted, My Discussion, To More Intimate Design Modulus, Although Elements, from My Skyscraper Design - Owe Many Major Influences to it - Although Always Intended, as a Secular Building - Sacred Buildings, Meant to Last Millennia - Tend to Incorporate, the Strongest Geometrical Elemental Forms, To Construct Them - Often First Similar Materials, Carved, Into Scale Models, in Antiquity - Where Physical Experiments, Could Be Subjected Upon Them - To Observe Stressor Limitations, They Could Withstand -

Another Characteristic, I have Found - Is Their Geographic/Geomantic Relationship, To the Solar and Lunar Cycles, Often Operating, Like Giant Sundials, Solstice and Equinox, Calendar - In Addition, To Providing, a Fixed Compass Position, of Static Observation, To Calculate, and Plot, The Cosmic Course, for the Ever Changing Movement of The Stars - Overall Design, For the Most Part, Pyramidical, Seems a Synthetic Attempt, To Emulate The Characteristics, of Holy Mountains - With Temple Complex, or Sacrificial Altar - Positioned, at the Energy Concentrating, Summit - The Piezo Electric Compression, Of Stacked and Staged Masonry, Often Filled With Quartz Striations, Traces of Radioactive Isotopes, Energetic Minerals - Existing, Under Downward Gravitational Compression - Increase Discernible, Exodynamic Magnetic Energy, Amplifying The Intensity, of the Rarified Space - Sacred Geometrical Structure - Where Synchronistic Phenomena, are Often Experienced - Due to Heightened Energy Formation - Both By Dynamic Piezo Electric, Pressured Building - And Psychological Valence Amplitude, Increasing In The Gestalt Field, Of Ecstatic Community, Who Utilize, The Sacred Space - A Sensorial Apperception, of Temporal Displacement Occurs, Felt Palpably, By Religious Ritual Adherents: Central Identifying Element, Of Synchronicity Phenomena: Dynamic Conjunction, Between Spiritual Mankind and Sacred Architecture Sacred Spaces, During Such Amalgamation, of Contributive, Synergistic Forces - Can Open Up a Stargate, in Extremes of Inducted Energy - Known in Pyramids, To Preserve And Mummify - Biological Substance, By Extension, Expands and Augments, Innate Spiritual Capacity - Like an Energetic Amplifier, Step Up Power Transformer - Striking Architectural Similarities, Between Many Ancient Cultures - Efficacy Brought to Bare - Even in Ruined Archeological State - By Simple Physical, Ergonomic Interaction - Within These Architectural Geometrical Structures: One Can Readily Conclude, When First Built, in Original Full Ritual Use - Their Sacred Geometry, Energy Generated, Was on a Much Higher Synergistic Magnitude, Accumulation Degree, of Productive Power Factor

Particularly, When Phases of the Moon and Sun, Solstice and Equinox: Calendrical Points, Comet and Meteor, Earth Passings - Would Additionally Influence the Power Demographic-Architectural Dynamic - To Greatest Synergistic Valence - Height Width

Ratios, Seem also Universally Consistent, in Terms of Relativity - Their Seems a Discreet Relationship, Optimal Paradigmatic, Architectural Scale, To Human Being - That Greatly increases, Energy Field Induction Efficiency - Existing Between Architect Geometer, and His Sacred Geometry, Upscaled Architecture, Approximately 100+/1, Vertical Ratio - Would Also Seem, Sacred Architecture, With Large Monolithic Stone Pyramids, With Marble Applied Panel Facia, Capped With a Metallic, Pinnacle Keystone Crown Point, Perhaps Alien in Origin, Technology Required, Lifting Giant Building Blocks, With Human Rampart Power, Levelated Monolith Tiers, Of Pyramid Dimensions, Was Lost in Time: Egyptian Civilization Moving Backwards, After an Initial Architectural Zenith, With The Valley of the Kings, As Retrograde Example - The Ziggurat Building, Built With Kiln Cured Clay Bricks, Easily Built Over Time, By Human Beings, Adds Credence, To This Construction Hypothesis - Living Organism, Bions, Spontaneously Form, Reichean Orgonomy, Created, by Repeated Endothermic, Exothermic Exposure Bombardment, Animating Clay Reaction, Complexity of Organic Compounds, Many Containing Minerals, Gemstones, Produce Piezo Electric Discharge, When Placed Under, Massive Gravitational Pressure, Babylonian, Brick Step Pyramid, Ziggurat

In Both The Pyramid and The Ziggurat - The Structure Acts, Like a Focusing Coil, Directing Energy, from the Ground Foundation, Axial Core, Upwards to the Apex Pinnacle - Internal Structures, such as the Kings Chamber - Built at Equidistant Center Of Internal Mass - Along Central Vertical, Axis Line of Great Pyramid - The Scientific Equivalent, of Heat Induction Oven - Recirculating Contained Energy, With Cyclonic Motion - Drawing Further Ambient Physical, Environmental Force, Unto Itself, Like Electro Magnetic Toroid Coil, Tidal Whirlpool, Mythological Charybdis Vortex - Given These Various Compositional Parameters - Categorically and Synergistically Contribute, To Purpose and Efficacy, of Sacred Geometry - I Would Seek, a Modern Modulus, New Original Paradigm, Another Semiotic Relationship, Architectural Vocabulary - To Integrate, the Best of the Past, Into an Aesthetic Parabola, and Fascinating Rubric, That Resonates, Unique Design Modernity - While Still Synthesizing, Sacred Space, By Correct Alignment, Geometric Formal Elements - Architected, Aligned Geomantic Cardinal Direction - Thus Creating Sacred Space -

Having Lived On The Navajo Reservation/First Nation, and on Hotevilla, Hopi Reservation, Third Mesa, Arizona - The Oldest City, In North America - I Was Made Aware, of the Kiva And Hogan, Indigenous Architecture, The Traditional Form, of Aboriginal Ritual, Sacred Geometry, Religious Architectural Space - The Kiva, A Circular Structure, Built Partially, Into The Central Summit, of Desert Mesa, Subterranean Clan Lodge, Where The Kachina, Prevailing Genius Loci Spirit, Dwells Within The Mesa Rock, There Venerated, Communicated With, In Religious Prayer and Ancient Song - Adding to Vast Codex, Accumulated Prophesies, The Hopi, Are Famously Noted For - The Separate Kivas, Entered Into, By Chosen Members, of Various Clan Societies, Through Ladder Access Point, In The Center, of Kiva Circular Roof Entrance - The Interior Space, Windowless, Illuminated Only, By Central Fireplace - During Yearly Summer Ceremonials - Each Traditional, Hopi Clan Society, Takes on the Regalia, and Costumed Affectation, Channels the Energy, En Masse, of the Pantheon of

Their Individual Kachina, Clan Ceremonials - That Are Hopi/Tewa/Zuni, Culturally Worshipped - The Line Procession Produces, Heightened Metaphysical Energy, Shared By The Assembled, Portions, of the Hopi Tribe, Witnessing In Participating Audience,

Traditional Hopi Families, Have Ceremonial Housing, Ancestrally Built, in the Traditional Adobe Brick, Clay Slip Slurry, Mud Applied Exterior, Multi-Stacked Architecture, Around The Central Plaza, Where the Kiva Structures, Are Centrally Located - The Roofs, of These Structures, Offer Additional Vantage Points, For The Elaborate Kachina, Summer Ceremonials - The Upswell of Metaphysical Energy, From The Mesa, Kachina Dressed Kiva Clans, Attending Tribal Audience, is Palpable and Real - Sometimes, Parts of the Ritual Costume, Often Fall Upon The Ground - Coveted by Culturally Ignorant, White Tourists, Are Known To Experience Poltergeist Activity, Upon Returning Home - Produced, From Such Stolen Ceremonial Objects, Usually Blessed Feathers, That Fall Off, Kachina Clan Costumes - Much Mail, Is Sent Back to the Hopi Nation Yearly, Returning These Objects, Ceremonial Kachina Items, With Apologetic Notes, Explaining The Poltergeist Phenomena, in Horrific Detail: Corresponding With The Objects, Being Brought Back, Into Non-Indian Homes, As Tourist Curiosities, And Collectors Items - With Negative Taboo, Metaphysical Consequences - These Paranormal Accounts, Are Well Documented, Testifying To Attending Metaphysical Phenomena, Protecting Mesa

Also Witnessing, Kachina Spirit Possession, Paranormal Phenomena, Several Dazed Adolescents, Falling Off Rocks Head First, As In Epilepsy Seizure, Or Fainting Spells, At The Peak of Hopi Mesa Ceremonials - In All Cases Picking Them Up, And Returning Them, To Their Homes, After Determining Their Families Location - Usually Met By The Mother, At The Door, Emotionless, In Similar Dissociated Psychological State - Taking, The Unconscious Child, From Me, Without Acknowledgement, Or Thanks, Offering No Additional Comment - Was Emotionally Perturbed, At These Experiences - Needless To Say, Respectfully Declined, Subsequent Further Offerings, Of Kiva Ritual Invitationals, Peyote Ceremony Participation: Having Seen, The "Trickster" Tewa Clown, The Hopi Mud Head, Kachina Spirits, in Remote Views, Living Within First Mesa, Near The Hopi-Tewa, Monolith "Navajo Kill Stone" - Additionally, Part of the Mesa Road, Spontaneously Collapsing - During That Summer Ritual Season, After Being Subjected, To High Intensity, Blustering Wind Dust Storms, Coming From The Direction, of Flagstaff, Like Saharan, Scirocco - The General Spiritual Tone, and Atmosphere, Fringing Upon, The Malevolent - This Further Verified, Ancestral Tribal Taboo Signs, Of Increased Witch Activity, In The Nearby, Box Canyon Systems, Located on the Contiguous Edges, Of the Hopi-Tewa, and Navajo Nations: No Mans Land, Unmarked Burial Ground, Ancestors Buried, Bound In Fetal Position - Providing The Corpse Material, and Infected Turquoise Jewelry, Mainstay Ingredients: Employed in Local Witchcraft, To Sicken Victims, Like Vietnamese, Punji Sticks - Historically Noted, in The Anthropological Literature, On the Academic Study, of The Tribes, in Question - Certain Clan Rituals, Like The, "Snake Dance", Were Also Hopi Tribe Banned, For Their Association With, "Left Hand Path" Witchcraft Practices, Made To Go Underground,

Into Black Magic, Ritual Secrecy - The Navajo Nation, Has Similar Cults - Far Less, Culturally Tolerated, When Tribally Ascertained, En Flagrant, Often Punishable By Summary Execution, and Death - Consanguine Couples, Generally The Perpetrators

Traditional Native American Shamanism, of Right Hand Path, Primarily Monotheistic In Creator God Theology, Has Been The Vanguard, Against Such Deleterious Effects, Of Negative Metaphysics, Upon Tribalized, Organized Societies - Positive, Generative Remedy, to Such Psychic Pandemic Disturbances - Sought Out, as a Means, of Spiritual Defense, Against Witchcraft Spell Casting, Poltergeist Activity, Psychic Attack, Sympathetic Magic Cursing Phenomena - The Traditionalist Shamans, Often Employ, Sacred Geometry Architecture, Their Spiritualized Medicinal Practices, Long Attested In Twenty Five, Thousand Years, of Trial and Error Empiricism, Employing the Unique Geomancy, of North American Continent, as Modus Operandi, To Create Their Scared Healing Structures, Build Upon Sacred Ground - It Is Through Their Discernment, Of Elemental Energy, Inducted and Focused, Into Sacred Space - That Spiritual Cure is Obtained, General Psychological Homeostasis, of Tribal Configuration, is Perpetually Sustained - These Systems of Architecture, in Slight Variation, Correspond To the Geometrical Matrixes, of Many Other Aboriginal Cultures - Cross Cultural Contact, Entirely Possible, in Such a Large Expanse, of Historical Time - Or Perhaps Morphogenic Resonance Prevailing, Inspiring These Diverse Cultures, Continents and Oceans Apart, To Independently Discover, For Themselves - The Replicating Sacred Geometry, Producing Rarified Spiritualized Space - The Shamanism Itself, Creates The Shaman: Peace and Tranquility, In Accord With Nature, Is What The Shaman, Strives For: Such Ubiquitous Nature, Is What The Shaman Draws Down, Cosmogony He Replicates, in Microcosm, Through Architecture, Prayer, Ritual Art and Sacred Song

The Navajo Nation, With Vast Territory, Which Encompasses, The Hopi Mesa System, Of Small Villages, is the Largest Indian Reservation/First Nation, Overlapping The "Four Corners" of Four Separate, American States - Arizona, Utah, Colorado, New Mexico - Containing Segments, of the Grand Canyon System, Monument Valley, Among Other Natural Wonders - Largely Desert Promontory, Covered With Sacred Sage, There Are System, of Holy Mountains, Sacred to the Navajo, Lesser Canyon Systems, Festooned With Petroglyphs, Cultural Remnants, Village Cliff Dwellings, of the Ancient Ones, the Anasazi: Legendary Ancestors, To Both Navajo and Hopi - Though Anthropologically Inconclusive - The Ancient Pre-Columbian, Indian Civilization, Original Architects, of the Elaborate Kiva Complexes, Like "The Castile", at Chaco Canyon, Devising Also, the First Sun Dial, in the Western Hemisphere, To Calculate the Solstice and the Equinox, Where Cistern Pools, Harvested Limited Yearly Rainfall, Desert Wash Arroyos Providing Flash Flood, Additional Water Sources, From Snow Topped Mountain, In Spring Time Thaw - Made Continuous Existence, in Harsh Desert Possible - These Mountain Systems, Became Sacred, To The Ancient Native Peoples, Integration Of Their Legends, Into Creationist Religion - Often Depicted, in Stone Relief Graffiti, Over Many Generations, Of Chronicled Rock, Carved Face Petroglyphs - Reconnoitered Pictographic, Recorded Prehistory, Extraterrestrial Visitations, of All Kinds: Fantastic Recollections, Of The Sky Gods, Helmeted Giants, Riding Variety of Unidentified Aircraft

The Mountains of Navajo, Contain High Amounts, of Uranium Ore, Energetic Sources, Of Radioactivity - Prime Locations, For the Navajo Shamans, To Establish Their Healing Practices, in Close Mountain Proximity - Octalinear Structures, Called Hogans, Log Cabin Constructed, Upon the Eight Compass Directions, Design/Built, to Wind Shelter The Sand Painted, Cosomological Healing Mandalas, Drawn Upon Earthen Floor - Elaborate Process, Requiring Many Hours of Work, is Specifically Designed, For The Ailing Patient, Is A Temporary Spiritual Artform, Ephemeral In Nature - The Patient Laid Down Upon It, Horizontally, Sung Over, In Ancient Tongue, To Bring The Patients Spirit, In Alignment, With the Great Spirit, Proper Alignment Essential, In Direction Based Curative System - Physical Proximity, to The Sacred Mountain, Inducts Environmental Ambient Energy, The Shaman Song, Like Resonant Vibrator, To Living Radioactive Rock, Echoes Produce, Amplification Effect - The Navajo Language Itself, Incredibly Complex, Like Basque, Finnish, Welsh, Romanian, No Known Linguistic Family Roots - "The Sing" Composed, of Such Remarkably Descriptive, Complex Language System, Further Enhances, The Energy Induction Processes - Sonically Resonating, With All The Possible, Phonetic Tones and Inflections, To Achieve Analogous Vibrational Reverberations, Within Living Mountain, Vibrating Radioactive Stone - Increasing Harmonics, Activate The Curative Power, Sung Over, The Direction Prone Hogan Insulated, Sand Painting Positioned, Prayed Over Patient, in Cascading Series - Of Geometrical Relationships, Scaled Down Ergonomics, Ritual Octalinear Hogan Space

The Navajo Reservation, Has The Largest Demographic Density, of Geodesic Domes, Perfect Structures, For Desert Conditions, Like the Trans-Siberian, Mongolian Yurt, and Eskimo/Inuit Igloo, Also Modern Variant, Upon Ancient, Inuit/Siberian/Mongolian/ Athabascan, Traditional Indigenous Form, of Tundra/Promontory, Sacred Geometrical Architectures, Like The Hogan - The Experience, On Navajo Reservation, Was a Synchronicity: Having Designed Similar Fusion Architecture, of Traditional Hogan and Modern Geodesic Dome, Years Before Visiting There - An Affirmation, of Inspired Designing Modulus - Found Sacred Geometry Parallels, Most Fascinating, Dwelling Philosophically Upon Them -  How My Original Architectural Design, Designated the Octalinear, Bore A Striking Similarity, and Intended Purpose, To Hogan Shaman Structure, Realizing Modern Fusion Design Purposeful, With Geodesic and Sacred Geometric Enmeshed, New Design Paradigm Possibility - Modern Configuration, of the Architectural Designs, The Two Most Prevalent, on Navajo Reservation - Purpose Built, For Ceremonial Healing - Built Upon Encompassed Proximal Coordinates, Near Sacred Radioactive Mountain Summits: 100+/1: Ratio, Vertical Scale, Like Egyptian Pyramid

Was Astounded Between, By Parallelism Of Design Philosophy - The Historical Efficacy, Of Sand Painting Healing, Energy Containment Field: Shaman Hogan - The Acoustical Engineering, of Octalinear Design, To Amplify and Increase Sound Fidelity, Navajo Shaman, Curative "Sing", Apparently Attributable, To Resonate and Channel, With the Radioactive Stone Mountain, in Close Echoing Proximity - To Draw Down The Healing Cosmological Energy - Color Sand Depicted, in High Detail, Drawn Upon the Hogan Floor - The Synergistic Use, of Natural Environmental Energy - Through Graduated

Dynamic Scale, Ergonomic Ratio of Hogan, Replicating the Geometrical Ratios, Of the Babylonian Ziggurat And Egyptian, Meso-American Pyramid, Existing Simultaneously, Upon Eastern and Western World Hemispheres - In Apparent Morphogenic Resonance

The Natural Formations of Holy Mountains, Generally Chosen, With Pyramidical, Geometric Peak Formation, Performing The Same Structural Attributes, in Architectural Sacred Ritual, Yet Much More Powerful: Being Made of Radioactive, Unquaried Living Stone - Akin to Holy Mountains of Sinai Desert, Turkish Anatolian Plane - Where The Biblical Prophets, Went to Covenant Prophesy, Directly Pray, To Their Creator God - Where Paranormal Phenomena, Would Spontaneously Manifest - Pantheons, Attributed To the Ancient Greeks and Romans, Mythology Existing, on Highest Mountain Summits, Also Usually Pyramidical, in Form and Function, Towering Highest Above The Clouds, Dwarfing Surrounding, Ancillary Mountain Chains - Warping Natural Forces, Into Itself -

Later, Living in Houston, Texas, in Rice University, Museum District, Visiting the Rothko Chapel, On The Grounds, of The De Menil Collection - I Was At Once Taken, By The Sense of Profound Tranquility, Upon Entering, the Rarified Space - The Subdued Rothko Painted Panels, Seemed A Catalyst, Energizing the Octalinear Chapel - Having Also Visiting, Mexico City, Shortly After, The Great Earthquake - Was Drawn, To The Main Catholic Cathedral, Within The Center of the Natural Disaster, Destroyed City - Upon Entering, Experiencing a Huge Amount, of Unleashed Metaphysical Energy, Radiating Up From the Cathedral Floor, Overpowering the Tremendous Vaulted Space - Later Ascertaining, The Main Aztec Pyramid, Was Archeologically Interred, Directly Beneath It, Seemingly Activated, By The Earthquake, Generating Enormous Free Energy - The Massive Cathedral, Became Orgone Accumulator, Energetic Reactor Containment Field - These Collective Experiences, Led Me To Believe, The Efficacy of Scared Geometry, Architectural Functionality, of Empirically Experienced, Sacred Space Was In Fact To Entrap, Natural Forces of Nature, Star Constellated, and Hyper Focused, Through the Complex Matrix Systems, of Fractal Geometry - Organic Passive Energy Reactor, Strike Point Vortex, For Focusing Free Energy Induction - Like Lazar Guiding Mirror Array, Reflective Matrixes, Entrapping Energy, In Labyrinthine Chamber

Designing My Structure, Using the Octagonal, Windows, Doors, Floor, Scale Graduated Renditions, Repetitions of Each Other - Blunted Pyramids, and Cubical Joints, Further Sophisticating The Matrix Complexity, of the Geometrical Architecture - Incorporating all The Sacred Geometrical Forms, I Had Encountered, In Architectural Odyssey - The Roof Keystone, A Quartz Crystal, Also Octalinear - Radiates Forth, Interlocking Geode Mosaic, Like the Navajo Sand Painting Cosmogony, of Traditionalist Shamanic Hogan - The Orientation, of the Entire Structure, Intended to Rotate, Upon Industrial Swing Bridge, Roller Bearings, Set Upon Circular Platform, Precise Compass Orientating, To True Magnetic North - Each Facia of the Octalinear Structure, Perpendicular to Cardinal Geomantic Force, Equidistant, In Each Cardinal Direction, Creating a Magnetic Center, Within Itself - A Sacred Space Intended, For Meditation: Like Sweatlodge, Yert, Tee Pee, Congregation Parliament, Long House of the Algonquin, and Viking Norse - All Those Accumulated Power Factors, Create Interdimensional, Synchronicity Phenomena Plane

The Fastening Systems, Windows, Emplacement of Tapered Angles, of Roof Elements - Created Dynamic Isotonic Tension, Like the Curvilinear, Ice Block Elements That Insulate and Fuse, Arctic Tundra Igloo - I Theorized Their Structural Similarity, Showed Cross Cultural Fertilization - Original Sacred Ritual Rarified Space, as the Primary Purpose - Directional and Geomantic Architectural Design Modulus Employed Living in Such Desolate Environment, of Nomadic Desolation - Psychic Ability Must be Honed and Amplified, To Locate the Migratory Herds, For Hunting - Agriculture in Such Northern Climes, Establishing Permanent Tribal Settlements, Impossible To Sustain - Intuitional Apperception, of "The Manitou", The Creator God, At One With Nature, in Religious/Shamanistic/Psychic Fusion, The Only Nomadic Hunting Method, To Survive

In Summation, Graduated Geomantic Ratios, in Relationship to Ergonomic Dynamic Relativity - Creates Synergy, in Rarified Sacred Space - Best Compounded into Complex Matrix Constellations - To Serve to Focus and Direct Heightened/Entrapped, Metaphysical, Inducted Energy - The Synthesized Architectural Space, Should Provide Resonant Cavity, Acoustical Attributes, With Comprised Structural Elements, Replicating In Identical Series, Analogous Placement, in Cartesian Coordinates - To Engender Modus Operandi, of Accumulated Energy, Animating Synchronicity, Producing Sacred Space, Articulated Through Geometry - Energizing Organized Human Ritual Activity Within it - Movement Dance Song Prayer, Meditation - Increased in Vibrational Intensity, Amplified by Resonant Cavity, Architectural Structure, Becomes Acoustical Chamber - Acoustical Engineering, a Discreet Holographic Sound Recording, Further Increases the Efficacy of Paranormal Phenomena, in Rarified Space Occurrence, Within Sacred Geometry Architecture - Engenders the Purification Process, Within the Architectural Structure, With Continuous Righteous Use Over Millennia Aeons, Energy Accumulator

Such Resonant, Frequency Generation, of Participating Human Adherent, Matched With The Vibration Tone, of Resonant Cavity, Architectural Matrix - Frequency Attuned, Sacred Geometric Structure - Synergistic Fusion, Between Architectural Space and Human being - An Extension and Induction Platform, of Reciprocating Energetic Action - The Resulting Dynamic, Causes Autonomous Space, Dream Like Timelessness and Overwhelming Power - Over Material Plane, of Mundane Consciousness - it is in this Contiguous Metaphysical Interdimension, Where Prophesy Comes, and Unobstructed Viewpoint, Born of Such Spiritual Intensity - That Allows Prophesy, to Paint Accurate Depictions, of Futurity - Remarkable In Undistorted, Pristine Electrifying Detail -

Sacred Song, Sonic Resonating Space, Eurythmy Dance, Where Wake of Dancing Molecules, Around Performance, Enlivens Dormant Vacuity - Fantastic Speeches, Rendered With Inspired Truth - Resonate Within, Complex Geometry of Stone - Leaving Permanent Record - Only Those Who Talk To The Stones, Can Glean Or Hear Dramatic Readings, of Sacred Liturgy, With Likened Faith, and Reverence For God Channeled Writ Will Actually Reactivate, the Spiritual Process, By Which It Was First Created, Speaking It Out, Energetically Reciprocating - Before Congregation Witnessing

Increased in Power, By Scared Geometry, Annunciating Before The Tabernacle -
Imbuing the Rarified Environment, with Amalgamated Levels, of Sacred Energy -
Synergizing, All Things Attuned, Within Itself: The Micro and Macrocosmos, Above, So
As Below - Equidistant, Within All Cardinal Directions, Compass Trued, To Polar
Rotational Axis, Shifting Magnetic North - Adjusted, as Required, To Compass
Synchronize, Base Rotating Octalinear Structure, Calibrating Directional Equilibrium -

The Generated and Induced Energy, is Not Dissipated, Continuing to Flow, Most
Mightily - This Appears, Constant Design Modulus, in Religious Architecture Throughout
The Ages, No Deviation, Is Entertained, Lest It Neutralize, It's Inherent Power - Intuition,
Discerned Calculation, Has Turned the Aesthetic, of Sacred Geometry Architecture,
Into Quantum Law of Metaphysics. Awareness Evolves, Creating Refinements, Sacred
Architectural Design Evolves - What We Feel, In Direct Metaphysical Experience, We
Often Transfer, Into Outward Manifestations - As The Unbroken Line Continues Onward,
To Unfold Before Us - The Architecture Evolves, Into Biomorphic Geodesic, Biotic Form,
Approximations, of Microscopic Structures, The Power Source of Life, From Single Cell
Biology, to Vast Universal Cosmogony - It Is In This Spirit, Offering Another Vantage
Point, Physical Perspective, Unifying Philosophy, Which Has Been, My Preoccupation,
To Do So - Segway, into General Fusion Physics, Likened To My Exploration, For the
Enochian Key, in the "Caduecuan Hieratic of Matrix Synchronicity" - My Matchmaking
Treatise, Upon Zodiacal Synastry Soulmating, Another Matrix System, That Produces,
Synchronistic Phenomena - Through Spontaneous, Community Dynamic Formation -
Flattened Hieratic of Astrological Pairings, That Create Synergistic Valence - Perfect
Synchronizing Mechanism, For Energizing Sacred Space, Religious Architecture -
Final Answer, to Jungian Transpersonal Psychology, Attempting To Break The Code,
In Metaphysical Question - What Tantric, Mantric, Magic Process, Brings Synchronicity
Forth, With Empirical Veracity - Palpable Sense of Timelessness, Heightened Psychic
Awareness, Highest Ascertained Degree, of Platonic and Romantic, Unconditional Love

The System of Fusion Physics, Presented Here, Yet Another Piece, to the Cosmic
Puzzle - An Abstraction, to the Mirrored Reflection, of Sacred Geometry - It Attempts To
Boldly Illustrate, The Physical/Metaphysical Periphery - The Interdimensional Plane,
Where Reality and Miracle Meet - Having Described Equally, in Scientific Parlance, As
Well As Laid Out, Fusion Physics Formula, With Classic Equations Listed, as
Philosophical Guidestones - It Proposes a, **General Field of Relativity**, In Which
Physical Phenomena, at Hyper-Energized State Behave - Like Synchronicity, Existing in
Simultaneity, Unconditioned by Spacio-Temporal, Conditioning - Energy Overcoming
Matter, At Discreet Moments, Miraculous Events - When Energy, Matter, and Space
Time, Form Perfect Storm, of Timelessness - Like a Theory of Rotating Star, Existing
And Not Existing, as a Black Hole Sun - An Interdimensional Portal, Realm of Antimatter
And Tachyon - We Accelerate To It, As It Decelerates To Us - The Threshold Of Light
Speed Broken, Everywhere and Nowhere, At Once - The Dreaming Mind of the Psychic,
The Prophet, Sage, Philosopher, Operates Much Like This - Denuded of the Physical
Body, The Astral Mind, Is Made Free To Travel, Back And Forth, In Time, Made Illusion
There - Viewing Reincarnation, Invention, Future Events, To Fuel Their Inspired Genius

To Reach, The Intended Goal - Many Are Those Who Attest, To This Process - The
Physical Mechanics, of Such Phenomena, Have Remained Elusive - Rather Than Delve
Into Metaphysical Speculation, I Would Rather Construct A Physical Model, To Further
Contemplate - Considering New Type Of Structure, Like The Skyscraper Design -
A Way To Reach New Heights, Of Understanding, Unswayed, Upon A Firm Foundation -
May It Add Well, to the Metaphysical Discourse, Upon Sacred Geometry, And Further
Inform It - Metaphysics, is the Science of Tomorrow, Alchemy, the Physics of Today -
I Would View Such Things, From the Holy Mountain Peak, Looking Into The Periphery

During Periods of Decline and Decadence, When the Transvaluation of Morals,
Becomes Bizarre Commonplace, Reaction Formation Takes Hold, The Philosophical
Become Targeted Individuals - One Must Chastise and Educate, The Forces That
Would Oppress Us, Often To No Avail - Such Convoluted Relationships, Essential
To Calm The Turbulence, Must Reckoned With Directly, To Expose Their Nonsensical
Nature - Like a Mirror Image of Reality, Distorting Two Dimensions, in Reverse:
Missing Altogether, The Empirical Verifying Third - All Profound Meaning Seems Absurd,
Waste of Time and Energy - Meanwhile All Things, of Value Disintegrate, Neglected
Fruit, Upon the Bower - Dissociating, From What Is Real, Illusion and Deception,
Hold Us, Death Gripped, In Their Thrall - Common Sense, The Path To Wisdom,
Spurned, and Cast Upon the Ground - Who There, Is Left To Pick Up The Broken
Pieces, Mend Antiquities Unvalued - Cast Off, Like Flotsam and Jetsam, From a
Sinking Ship of Fools - Pirates on the Horizon, See The Floating Trail - Like Chum
In Water, Attracting Such Schools of Sharks - They Circle Round, The Injured Prey -
Frenzied, By The Smell of Blood, Remorse is Alien, To Their Predator Sensibility
Nothing Living, Will They Leave Behind - Cannibalizing Each Other, If Obstructed
From The Feast - Meanwhile the Watchers, Stop and Stare, Too Far Away, From
Promiscuous Course, Offering Silent Condolence, As They Turn Away - Having
Warned the Wayward, and the Wicked Before: Supplied Them With Instruments,
Of Navigation, Thrown Overboard, in Submerging Frenzy - Found Useless,
In The Uneducated Hands, of Post Technological Idiots, And So The Story Goes -
Perhaps This Story Resonates, Or Falls Upon Deaf Ears - Some See The Signs,
Like a Cracking Dam, Some Leave the Valley For Higher Ground, Other Hear The
Concrete Crumbling, Thinking it Lightning, And Fall Back To Sleep - Others, Oblivious
Do Not Awaken, The Downpour, In Deluge Continues, The Dam Gives Way, Under
Unbearable Pressure, The Valley Returns to Lake, Many Lost Souls, Are Taken -
Many Are The Signs, That Would Deter Us, From Such Calamity - Listen For Them,
Do Not Hesitate, To Act Intuitively - Ascend To Higher Ground, Save Yourself -
Even The Most Clever, Often Fool Themselves, Delusions of Grandeur, Often
Obscure, The Path Not Taken, In The End We Must Meet Our Maker, Think Well Upon It
Trespass Not, Upon, Your Fellow Man - Often They Are Disguised, in Borrowed
Garments, Sent To Test the Wicked, of the World - They, Like The Dam, Have Their
Breaking Point - Like the Tao, Nature Remains, At Their Beck, and Call - Such Power
Of Fate, Remains All Around Us: Stay Close To The Center, Deviate Not Upon The Path
ELOHIM Pahana @ Hopi, Katimavik, Gathering The "Red Ones", Golden Alchemic Fire:
"Lost White Brothers": Translated Saints, African "Mzungu", Cause World Paradigm Shift

# GENERAL THEORY OF FUSION PHYSICS:

## VISIT: www.Brooklynrail.net

1830, **"Planet Type"**, Steam Locomotive, Magnetometer Discovered, **"Displaced Archeology"** Within **"Atlantic Avenue Tunnel"** Discovered, R. Diamond, 1980

257

1) PURE OUTER SPACE ENVIRONMENTALLY CONDITIONED AS A HERMETICALLY SEALED VACUUM – TEMPERATURE REGULATED @ ABSOLUTE ZERO – IS THE PEREFCT INDUCTION MEDIUM FOR DARK MATTER – A SYNTHESIZED CONDITION EMPRICALLY REPRODUCED WILL YIELD ZERO POINT ENERGY ON AN INFINITE QUANTUM LEVEL - AND CONSEQUENTLY IS THE ONLY GRAVITY REPULSING SYSTEM THAT IS SUFICIENTLY POWERFUL ENOUGH TO ACHIEVE LIGHT WARP SPEED FOR FEASIBLE INTERGALACTIC SPACE TRAVEL/ THROUGH QUANTUM LEAPING THOUGH WHITE AND BLACK HOLES – CREATING SPECIFIC HIGH ENERGY INTERDIMENSIONAL ATEMPORAL NAVIGATIONAL PATHWAYS, SIMULTANEATY VIADUCTS, THROUGH THE TIME SPACE CONTINUUM.

2) WHITE HOLE/ BLACK HOLE ZERO POINT ENERGY PRODUCTION OCCURS BY SUPER COMPRESSION OF MATTER BY GRAVITATIONAL IMPLOSION @ AT LOW TO VIRTUALLY NO RPM'S – OR BY HYPER ACCELERATION OF UNCOMPRESSED MATTER @ HYPER ACCELERATION/ INCREASING MASS INFINITELY TO 10 TO THE 94TH POWER/ CM CUBED. CREATING BLACK HOLE SPECIFIC ANTI-GRAVITY PHENOMENA – DE-ACCLERATION CAUSES AN OBJECT TO RE-EMERGE INTO THREE DIMENSIONAL SPATIO-TEMPORALLY CONDITIONED SPACE THROUGH A WHITE HOLE – ZERO POINT ENERGY CAN ALSO BE INDUCED BY THE HYBRIDING OF THE TWO PHYSICAL CONDITIONS – DE-ACCLERATION CAN ALSO BE ACHIEVED BY COMPRIMISING OR BREACHING THE VACUUM STATE – AND/ OR INCREASING THE ENVIRONMENTAL TEMPERATURE – THEREFORE INDUCING THERMODYNAMIC ENTROPHIC BREAKDOWN INTO SUBLIGHT THREE DIMENSIONAL QUANTUM MECHANICS RANGE.

3) ZERO POINT ENERGY IS INCHOHERANT AND NON-SENSIBLE/NON-GAUGEABLE – OMNIPRESENT IN AN INFINITELY EXPANDING QUANTUM FIELD = THE UNIVERSE, COMPRISING INFINITY WITHIN ETERNITY.

4) DETECTION: MEASURING ZERO POINT ENERGY/ FREQUENCY GENERATION IS DIFFICULT – DUE TO UNIFORM FIELD – WITH PERCEPTIBLE DIFFERENTIATION ONLY OCCURING IN SPACIO-TEMPORAL/ FREQUENCY SHIFT ANOMALIES (DOPPLER SHIFT) EXISTING IN UNIQUE QUANTUM CONDITIONED SPACE PHENOMENA (DISCREET ANOMALOUS QUANTUM EVENTS)

5) ORTHOGONIC = VIRTUAL INTERDIMENSIONAL ORTHOGONIC FLOW/FLUX CONTINUUM OF THE DISCREET QUANTUM INDICATES A WHITE/BLACK HOLE REFLUX THIRD AND FOURTH DIMENSIONAL QUANTUM DYNAMIC/ ENERGETIC/ SYNERGISTIC INTERFACE = CREATING A PERPETUAL BREEDER REACTOR FOR MATTER/ANTI-MATTER IN AN INFINITELY EXPANDING UNIVERSE.

6) ZERO POINT ENERGY EXISTS IN A FREE RADICALIZED FORM, IN A UBIQUITOUS HYPER-AGITATTED – EXCITATIOUS – INTER-DIMENSIONAL STATE.

7) THE ZERO POINT ENERGY REFLUX (WHITE HOLE/ BLACK HOLE) WHICH EXISTS AT A ULTRA HIGH FREQUENCY LEVEL – DOES NOT READILY INTERACT WITH THREE DIMENSIONALLY CONDITIONED MATTER – UNTIL IT IS STEPPED DOWN SUFFICIENTLY TO SUBLIGHT BAND WIDTH AND QUANTUM – EFFECTIVELY DE-POTENTIATING INCIPIENT ENERGY, ERGO – "CHOPPING" ITS FULL POTENTIATED BAND WIDTH AND INHERANT INDUCTABLE POWER FACTOR POTENTIATION.

8) PHYSICAL MATTER MUST BE HYPER-ACCELERATED TO INTERACT WITH ZERO POINT ENERGY (TO ACHIEVE THE DARK MATTER INDUCTION POINT @ THE HIGHEST CO-EFFICIENT) THIS CAN BE ACHIEVED THROUGH PROPULSION – ROTATION – THIS CAN ALSO BE ACHIEVED BY HIGH VOLTAGE IONIC FLOW @ LIGHT SPEED POWER FACTOR/ AMPLIFIED BY TRUNCACTED ARRAYS OF FOCUSING COILS – AND/OR BY IONIC FLOW THROUGH SUPER CONDUCTIVE PLASMA FIELDS = DE-POTENTIATED THERMODYNAMIC ENTROPHIC BREAKDOWN = BY SUSTAINING THE REACTION IN SUPER COOLED ABSOLUTE ZERO – HERMETICALLY SEALED VACCUM SEALED REACTION

CHAMBER / AND/OR THROUGH FOCUSED PARTICLE BEAM ACCLERATION/ GENERATED AND AMPLIFIED THROUGH GEODISICALLY ARCHITECTED RESONANT CAVITIES = AFORESAID LAZAR APPARATUS SYNERGISTICALLY ACTING UPON INCIPIENT MATTER – AMPLIFIED BY THE GEODISIC MATRIX WILL EFFICIENTLY PRODUCE ZERO POINT ENERGY, WITH A PREVIOUSLY THREE DIMENSIONAL FIELD.

9) LAW OF ZERO POINT ENERGY = HIGH FREQUENCY ZERO POINT ENERGY DOES NOT INTERACT WITH THREE DIMENSIONAL MATTER – UNLESS THE MATTER IS RAISED UP TO THE ULTRA HIGH FREQUENCY BAND WIDTH – THUS FOCUSING AND INDUCTING AFORESAID ENERGY (DARK MATTER) BY EXPONENTIALLY INCREASING THE VELOCITY OF THE MATTER TO LIGHT WARP SPEED – MATTER WILL QUANTUM LEAP FROM 134,000 MPS TO 186,000 MPS OCCURS IN SIMULTANAETY

10) CONICAL TESLA (FOCUSING COILS) IN A STEAM VORTEX WILL CREATE BALL LIGHTNING INDUCING A SCALAR EFFECT AT SEA LEVEL ACCOMPLISHING A FORM OF DARK MATTER INDUCTION – THE EARTH ITSELF FUNCTIONING AS AN OPEN DELTA 33 DEGREES ROCKING TORROID – EMPIRICALLY ESTABLISHING A MACROCOSMIC PLANETOIDAL QUANTUM MODEL TO EMPIRICALLY REPLICATE AND ENERGETICALLY HARNESS.

11) A CADUECEUS COIL MANUFACTURED WITH PERFECTLY BALANCED COUNTERWINDINGS – EQUALATERALLY OPPOSING NUETRALIZED WINDINGS OVER A FERRITE CORE – HYPERACTIVE ENERGY POTENTIATED TO ZERO POINT IN A VACUUM CONDITION AT ABSOLUTE NEGATIVE ZERO DEGREES CELCIUS.

12) POLARIZED VACUUM/ WITH INTERNAL MAGNETIC BUCKING (HIGH FREQUENCY GENERATION/ MODULATION) = AETHER/DARK MATTER THROUGH CADUECEUS COIL

13) "FLYWHEEL" = ZERO POINT ENERGY = HIGH FREQUENCY MAGNETIC BUCKING FIELDS IN HERMETICALLY SEALED VACUUM CHAMBER @ NEGATIVE ZERO DEGREES – EQUIPPED WITH MAGNETICALLY SUSTAINED AXIS ANTI-GRAVITATIONAL BEARINGS.

14) CREATING SPECIFIC ANTI-GRAVITATIONAL/ LEVITATIONAL – ELECTROMAGNETIC/ DARK MATTER INDUCTION FIELD = EMPIRICALLY RE-CREATING PLANETOIDAL/GEO/ASTRO-PHYSICAL = MICRO-REPLICATION OF UNIVERSAL FIELD THEORY.

15) INFINITE ACCELERATION OF MATTER INDUCTED INTO DARK ENERGY THROUGH HIGH FREQUENCY PERMITIVITY WAVES PULLING THE OBJECT TOWARDS THE INTERDIMENSIONAL COMPONENTS OF PROFOUND ELECTRO MAGNETISM – CYCLOTRON = INTERCEPTOR = PERPETUAL ENGINE/GENERATOR = WARP DRIVE SPACE PROPULSION SYSTEM = TIME DISPLACEMENT DEVICE = HIGH ENERGY CADUECEUS INDUCTION COIL.

16) NON-LINEAR HYDRO-DYNAMIC AND/OR ELECTRODYNAMIC PARADIGM COHERANCE ESTABLISHED FROM RANDOM TURBULENCE = FORMING PLASMOID VIRTUAL COHERANT – SYNERGISTICALLY BALLANCED – MACRO/MICRO SUBSTRUCTURE = RECONFIGURED OUT OF ENTROPHIC THERMODYNAMIC BREAKDOWN AND REORDERING OF THE THERMODYNAMIC SUBLIGHT CHAOS.

17) LORENTZ CONSTANT INDICATES MATTER RESONATES WITH HIGH ACCELERATION AMPLIFYING DENSITY TO INFINITY – MATTER TRANSFORMATES INTO ANTI-MATTER = INDUCING LIGHT WARPING AND ANTI-GRAVITATIONAL FORCE = TIME DISPLACEMENT FIELD.

18) THE SECOND LAW OF THERMODYNAMICS WAS MODIFIED BY ILYA PRIGONINE (NOBEL LAUREATE 77' CHEMISTRY) = REVISION/ REFUTATION OF "CHAOS THEORY" ALL MATTER/ ENERGY EVENTUALLY RE-ORGANIZES ITSELF AFTER THERMODYNAMIC ENTROPHIC BREAKDOWN RE-ESTABLISHING RE-STABILIZATION IN A NUETRALIZED HIGH FREQUENCY INTERDIMENSIONAL

QUANTUM SPACE = EVENTUALLY RE-ESTABLISHING COHERANT ORDER/ BEYOND ENTROPHIC THERMODYNAMIC BREAKDOWN/ PLASMOIDAL VORTEX/ PROGRESSIOANL ROTATION (A HELLICAL ROTATION AROUND A FERRITE CYLINDER OR A PLASMOIDAL E/M AXIS = FERRITE THAT TRUNCATES INTO A TORROID IS THE OPTIMAL DESIGN PARADIGM = THE PHYSICS FORMULA LOSELY FOLLOWS THE PATH OF THIS VERBAL EXPLAINATION OF ZERO POINT ENERGY/NUCLEAR COLD FUSION REACTION.

19) TACHYON FIELD QUANTUM CONDITIONED STRING FLUX SPECIAL EVENT, ANTICIPATES "BLACK HOLE HORIZONS" – AS TACHYON PARTCILES DE-CELLERATE INTO SUB-LIGHT MATERIALITY, FROM OUT THE FOURTH TO THIRD DIMENSIONAL STATE: BREACHING HYPERSPACE IN A PLASMOID ENERGY TEAR, ANALOGOUS TO MODELS OF THE BIG BANG THEORY – CONVERTING MOMENTARILY, ANTI-MATTER INTO MATTER: CONDENSING INCIPIENT LOCALIZED SPACE INDUCTED, ZERO POINT ENERGY, INTO AN ANTI-GRAVITATIONAL FOCAL POINT OF DARK MATTER; CREATING SUBSEQUENT QUANTUM AND SPACIO-TEMPORAL DISPLACEMENT, PROXIMAL NON-LINEAR ANOMALIES – WHICH CAN BE TECHNOLOGICALLY SYNTHESIZED, HARNESSED AND EXPLOITED, TO CONTROL AND NAVIGATE THE TIME SPACE CONTINUUM, TELEPORTATIONALLY, IN A FOUR DIMENSIONALLY CONDITIONED QUANTUM LEAP: THIS OCCURING BETWEEN THE NEAREST PAIR OF TWIN MATCHED WHITE AND BLACK HOLE AXIS FORMATIONS = PRODUCING EMPIRICALLY, SIMULTANETY AT AXIS "C"

20) NUCLEAR FUSION PHYSICS, REPLICATES SOLAR ASTROPHYSICS, ON A MACROCOSMIC LEVEL – PRODUCING A GRAVITATIONAL FIELD, WHICH WHEN HYPER-ENERGIZED, WILL FOLD INTO IT'S OWN DISCRETE RELATIVE TIME SPACE – INTO TACHYON PRODUCTIVE FOUR DIMENSIONAL ANTI-MATTER, MATRIX SUSTAINED IN SIMULTANAETY, AS A BLACK/WHITE HOLE QUANTUM EVENT.

21) HYDROGEN, IN THE RARE ORGANIC MOLECULAR FORM OF DUETERIUM , IS A BASIC FUEL OF COLD FUSION NUCLEAR PHYSICS: WHEN ATOMICALLY FUSED WITH PLUTONIUM, WHILE BEING STRUCK BY A LINEAR ACCELERATED/CYCLOTRON ACTUATED PARTICLE OF RADIOACTIVE TRITIUM ISOTOPE – WILL IGNITE A NUCLEAR FIRE, TWELVE MAGNITUDES HOTTER THAT THE INTERNAL CORE OF THE SUN: COLD FUSION REACTION, AS A POWER SOURCE OF TEMPORAL DISPLACEMENT TECHNOLOGY, IS CRITICAL IN PRODUCING A SUSTAINED AND CONTOLLED BLACK HOLE/WHITE HOLE, ZERO POINT INDUCTION EVENT.

22) DUETERIUM, A RARE NATURALLY OCCURING RADICAL HYDROGEN ISOTOPE – CAN BE FOUND AND FARMED FROM THE STILL WATERS, OF SWAMPS AND FYORDS. DUETERIUM CAN ALSO BE SYNTHESIZED, BY EMPLOYING BREEDER REACTORS, ATTACHED TO A COLD FUSION REACTOR, SLUSH HYDROGEN PLASMA CORE: THE SLUSH HYDROGEN TORROID SUPERCONDUCTOR MAGNETIC RING, AT THE CENTER OF THE COLD FUSION NUCLEAR PILE – THE MAGNETIC TORROID FILLED WITH RADIOACTIVE HYDROGEN SLUSH, CAN BE TAPPED AND PUMPED OUT OF THE CENTRAL TORROID RING OF THE FUSION REACTOR – AND THEN ALLOWED TO HEAT UP, FROM THE EN VACUO CONDITIONED STATE OF ABSOLUTE ZERO, WITHIN THE REACTOR CORE, WHICH SIMULATES THE HEATLESS VACUUM OF DEEP SPACE – ONCE REMOVED FROM THE REACTOR CORE, THE HYDROGEN SLUSH RETURNS TO A GASEOUS STATE, AND SPONTANEOUSLY BEGINS TO RELEASE RADIOACTIVATED HYDROGEN PARTICLES, INTO A WARMED EXTERNAL ENVIRONMENT – SAID NUCLEAR FALLOUT CAN BE TRAPPED AND TARGETED INTO A PRESSURIZED TANK OF DISTILLED WATER – DESIGNED LIKED A NATURAL GAS TOWER, PRESSURIZED BY A SYMETRICAL RINGED ARRAY OF HYDROLIC PRESSES, COMPUTER MONITORED TO HYDRODYNAMICAL SUSTAIN A PRESSURIZED EQUALIBRIUM, EQUIVELENT TO A THOUSAND FEET OF STANDING WATER, OR A HUNDRED NAUTICAL ATMOSPHERES OF HYDRAULIC WATER PRESSURE, THE DISTILLED AND PRESSURIZED WATER IS FURTHER REFRIGERATED TO A POINT JUST ABOVE FREEZING, TO REPLICATE THE ENVIRONMENTAL CONDITIONS OF RIVER BOTTOM STILLED FYORD WATER: THE BOMBARDMENT OF SAID STILLED WATER, BY THE GAS RELEASED MOLECULAR STREAM OF

RADIOACTIVE HYDROGEN PARTICLES, WILL SYNTHESIZE ENOUGH DUETERIUM, TO PRODUCE A PERPETUAL ENERGETIC NUCLEAR CYCLE, SIMILAR TO THE LIFE AND DEATH OF STARS: RECYCLED NUCLEAR HYDROGEN SLUSH FALLOUT, PROVIDING AN ENDLESS SUPPLY OF SYNTHETIC DUETERIUM FUEL, FOR PERPETUAL COLD FUSION REACTION.

23) THE STILL WATERS OF BOTH DEEP RIVER FYORDS, AND INLAND SWAMPS, UNAFFECTED BY COASTAL TIDAL ACTION, BOTH BOMBARDED BY LOW LEVEL PHOTONIC NUCLEAR RADIATION OF THE SUN, ACOMPLISHES THE SAME TASK TO PRODUCE ORGANIC DUETERIUM, OVER THOUSANDS OF YEARS: THE STILL WATER MOLECULES NEVER CHANGING PHYSICAL POSITION, IRRADIATED OVER ENORMOUS PERIODS OF TIME, WILL MOLECULARLY BOND AND MUTATE INTO RADIOACTIVE ISOTOPES, PHOTOSYNTHESIZED INSITU BY THE NUCLEAR RADIATION OF THE SUN: A FUSION REACTOR, ESSENTIALLY A MANMADE VERSION OF A HYDROGEN STAR, CAN CREATE THE SAME MUTATING PARTICLE REACTION WITH IT'S OWN NUCLEAR FALLOUT, ON STILLED WATER, INSTANTAEOUSLY, UNDER CONTROLLED LABORATORY CONDITIONS

24) ANOTHER POTENTIAL METHOD OF PRODUCING HEAVY WATER/DUETERIUM, IS TO BOMBARD A PRESSURIZED STASIS TANK, OF PURE $H_2O$, PRESSURIZED AND CHILLED TO A HUNDRED ATMOSPHERES, WITH FROZEN PARTICLES OF RADIOACTIVE HYDROGEN, SHOT THROUGH A FORTY TWO INCH, CENTRIPICAL LINEAR PARTICLE ACCELERATOR, INTO A GUASS SHIELDED AND LEAD LINED, TARGETING FACETED BOMBARDMENT CHAMBER – THE ACCELERATOR SHOT PARTICLE OF RADIOACTIVE HYDROGEN, DE-CELLERATING FROM NEAR THE SPEED OF LIGHT, WILL EVENTUALLY BOND IT'S FREE RADICAL HYDROGEN PATICLE, TO FORM HEAVY WATER, WITH THE DISTILLED AN PRESSURIZED WATER MOLECULES WITHIN THE TANK, FORMING SYNTHETIC NUCLEAR GRADE DUETERIUM, WHICH CAN BE SUBSEQUENTLY USED AS NEW ENDLESS SUPPLY OF FUEL FOR PERPETUAL ZERO POINT INDUCTION FUSION REACTOR, IN EFFECT CREATING A SYNTHETIC NUCLEAR SUN.

## "SKY BELL":
### ANTI-GRAVITATIONAL, HYDROGEN PLASMA, NUCLEAR FUSION, PERPETUAL ENGINE

Bottom View     Side Cut-Thru

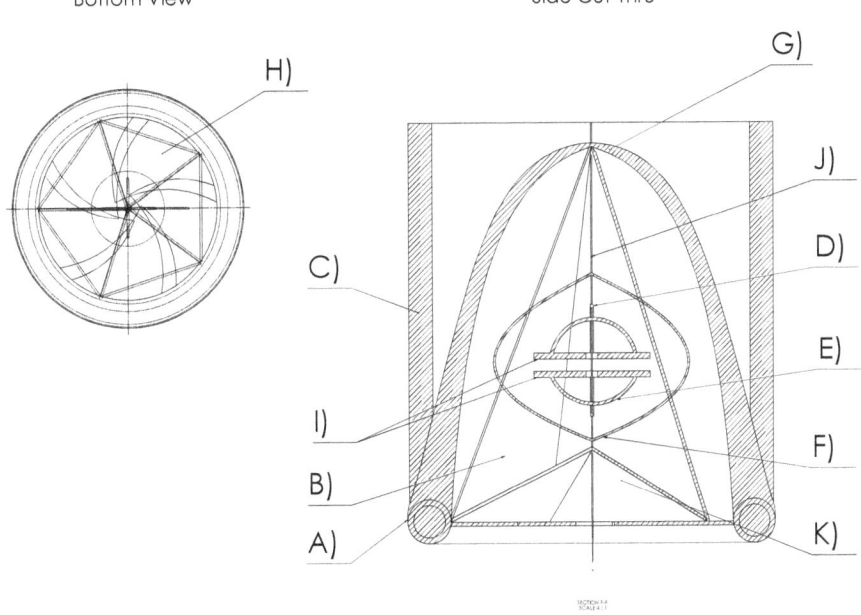

A) "Plasmatic Slush Hydrogen" Anti-Gravitational Toroid, "Floating Ring" (Superconductive Nuclear, "Chemical Plasma Flywheel")

B) Inverted "Jacob's Ladder" Protracted Compass, "Spline Sets", for "Triple Frequency" SONAR, 'Staggered Transmission' (UHF, VHF, "Linear Accelerated": Lazar Photon, Reactor, Toroid Bombardment: Triple Oscillation, Frequency Regulator) (Used as Circulatory/Vibrational Pump Mechanism, Frequency Agitating, Fusion Reactor, Dark Energy, Induction Vortex, for "German Bell Type, Hydrogen Plasma, Chemical Toroid, Nuclear Fusion, Superconductive Chamber") -

C) External "Double Hull" Reactor Shell, Domed Bell Cylindrical Encapsulation, W/Gausing "Magnetic Bottle Shielding"

D) "Upper Central Axis"

E) "Lower Center Axis"

F) "Dark Energy, External Space, "Zero Point Energy", Induction Rod"

G) "Upper Domed Bell" Tapered Reservoir, Uranium U235, Deuterium: Heavy Water, (2H, D)

H) "Sky Iris", W/Gause Shielded, Adjustable "Space Portal" External Environmental Space Aperture

I) "Magnetic Isolation Transformer", Twin Polarized, Electro-Magnetic Discs

J) "Main Axis", Adjustable Disc Transformer, "Frequency Fine Tuning" Central Devise, Oscillation Calibrator, Field Configuration, Intersecting Support Armature Array

K) "Jacob's Ladder" Array, Intersecting Support, "Base Elements"

1). $E=MC^2 = A = B \times M \times ACC \times C =$ SIMULTANEITY

2). $@AXIS(C) \equiv (S = \frac{\infty}{\infty} = \theta C \sqrt{\frac{\lambda_0}{0^\circ}}) = (A^2 + B^2)$

3). $A/G\infty/ACC/@/OR\ A@\ 90°@\ OR\ \infty\ ACC = \infty$

4). $136 < |\alpha| < 137 = P = \frac{MV^2}{2} @ \infty ACC = \alpha$

5). $\alpha < 136 = \infty ACC = \alpha = \infty = 186,000$

6). $MP < 137 = 134,000\ MPS = P = \frac{MV^2}{2}$

7). $\equiv$ (QUANTUM LEAP) $+ 52,000$

8). MPH [QUANTUM LEAP OCCURS

9). IN SIMULTANEITY @ AXIS(C)

10). $= 136 < 186,000\ MPS < =$

11). FACTOR) POSITIVE POWER

12). SIMULTANAEITY @ AXIS(C) =

13). $\{V = P = \frac{MV^2}{2} \sqrt{\frac{e}{P \times M^2}}\}\ \lambda = P(w) = \frac{hw^3}{2\pi^2 C^3}$
(MOMENTUM/VELOCITY SQUARED) (ELASTICITY DIVIDED BY DENSITY MASS SQUARED) (ZERO POINT DENSITY FUNCTION)

14). $< \infty ACC // C_i = C_0 \left(\frac{wp-}{wp+}\right)^2 // \lambda_0 = C_0 = \epsilon_0 \frac{S}{D}$
(RESONANT WAVE AMPLIFICATION) (ELECTRO VACUUM CAPACITANCE)

15). $= \infty ACC = 10^{94}\ GRAMS/CM^3 = 136 < |\alpha| < 137 = \infty$
(VALENCE POTENTIAL @ ZERO POINT ENERGY) (INDUCTION AMALGUM FORMULA)

16). $\equiv \{V = f \cdot \lambda < \frac{\lambda}{\infty} = \frac{V\infty}{f\infty}\}\ \infty = \overline{134,000\ MP < 186,000\ MPS}$
(LIGHT WARP ACCELERATION FORMULA) (INF SPACE CREATED BY INF VEL/FREQ) (<= 52,000 MPH QUANTUM LEAP + 36% POWER FACTOR)

17). $\equiv P(w) \frac{hw^3}{2\pi^2 C^3} < \infty_{ACC} = \sum 136 < |\alpha| < 137$
(BOYER/LORENTZ INVARIANT) (GENERAL RELATIVITY)

18). [ZERO POINT DENSITY FUNCTION]
$\oint \cdot PV^2 = \int \frac{10^{94}\ GRAMS/CM^3}{10^{1/26}(\Omega \cdot J)} \times \frac{\theta}{I\infty} \cdot \frac{E_0}{(\pi^2 \cdot \infty ACC)} < \frac{a_0 \cdot \oplus}{}$
(MAGNETIC VACUUM BUCKING EQUATION)

271.B.3

19). $= \{\omega p \theta \sqrt{\frac{NQ^2}{\epsilon_0 m}} \times \vec{A}_0 = \frac{L}{\text{(Faraday Induction)}} = \frac{dm}{SNQ^2}$ (Carrier Charge / Carrier Mass) / (Electro Separation / Electro Area)

(Theta Wave Plasma Freq) (Vacuum)

20). $C_0 = \epsilon_0 \frac{S}{D} = \infty = \theta\omega = \sqrt{\frac{1}{L(C_0 + C_1)}} =$
(Theta Wave) (Circuit Resonation) →

21). $\sqrt{\frac{SNQ^2}{dm(C_0+C_1)}} = \sqrt{\frac{NQ^2}{\epsilon_0 m(1+C_1/C_0)}}$
(Circuit Resonation) →

22). $= \frac{\omega p - (\theta)}{\sqrt{1+C_1C_0}}\} = \frac{\vec{A}_0}{T=0°}\sqrt{\frac{P = \frac{MV^2}{2}}{10^{12}}} / \frac{10^{94} \text{GRAMS/CM}^3}{= A/G = \underline{\text{LEVITATION}}}$
(Anti-Gravity Function)

23). $\equiv \omega p \theta \times \pi^\infty \equiv 10^{94} \text{GRAMS/CM}^3 \cong$
(Flywheel Acceleration Formula)

24). $(\sum \frac{1}{2} h\nu) = \sum \partial^2 Q/\partial x a^2 = \mu^2 Q$
(Infinite Zero Point Energy) (Schrödinger Scalar Wave Equation)

25). $\sum = evr/c = mv^2/H = r = mcv/eH$
(Average Magnetic Moment) (Charge/Mass Particle)

26). $\int_\infty^t \propto \int_{0°}^\infty = \psi \vec{A}_0 \psi(X) = \sum_{\alpha 136}^{137} \varphi \cdot \psi$ (cont)

27). $\underbrace{136 < |\alpha| < 137}_{\text{(Amalgum Equation/Quantum Leap)}} \sum_{\alpha 136}^{137} \varphi \cdot \psi \vec{A}_0 = \sqrt{\frac{\omega^3}{\theta}} \cdot \frac{\vec{A}_3}{\text{(Free Radicalized Vacuum State)}} \sqrt{\frac{10^{12}}{10^{94}}} \frac{A/G = \text{LEVITATION}}{\text{GRAMS/CM}^3}$

28). $\sum\int\equiv \triangle^H_{(\phi \cdot \Delta \frac{33°}{130°})} \equiv [\partial m = \frac{-0°}{T} = \frac{-0°}{T}\int_{A_4}] \equiv (\triangle 33°\cdot 3\cdot 2)(\triangle)$
(Open Delta) (E/M Mass) (Annihilation of Thermodynamics) (Open Delta Induction Array) (Open Delta Toroid in Caduceus Coil)
(4 Dimensional State SE Scalar Wave Integral) (Prime Mover/Toroid Arrays)

29). $\int_E^M \} = \vec{A}_0 \sqrt{\frac{\psi \mu \nu}{E = h\nu = h/\lambda}} \overset{\text{(Annihilation of Lambda Through Velocity)}}{=} \rho(T, \vec{A} \infty)$
(Energy/Mass Integral Function) (Zero Point Energy) (Planck's Constant) (De Broglie's Constant)

30). $\sum_{N=0}^\infty e \vec{A}_{N_Z N}(T) = P = \frac{MV^2}{2} \sqrt{\vec{A}_0 / T_0 \frac{(-T)}{(-T)}} \frac{\int_\infty^t \propto \int_{0°}^\infty}{\overline{136 < |\alpha| < 137}} (A)$
(Anti-Gravity/Anti-Matter Unified Field Equation)

31).
32). $\equiv \sum_{\alpha 136}^{137} \pi^\infty \xi_0^\infty \sqrt{G(10^{94}/CM^3) \cdot P = \frac{MV^2}{2}} = \int \partial_{\vec{A}_4} = \vec{A}_0 = \infty$ (Page Number 271, B4)
(Time Displacement Theory)

31).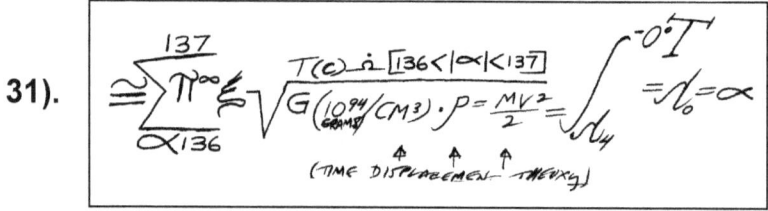

32).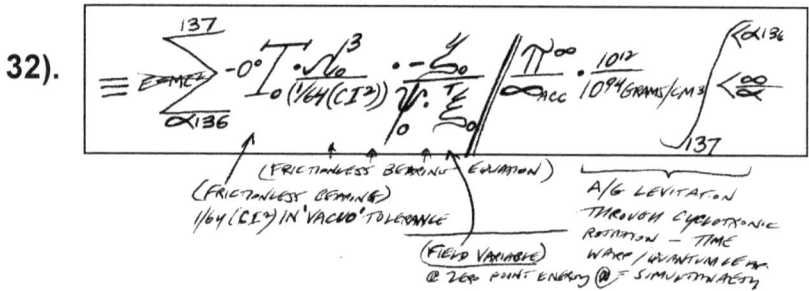

18.5).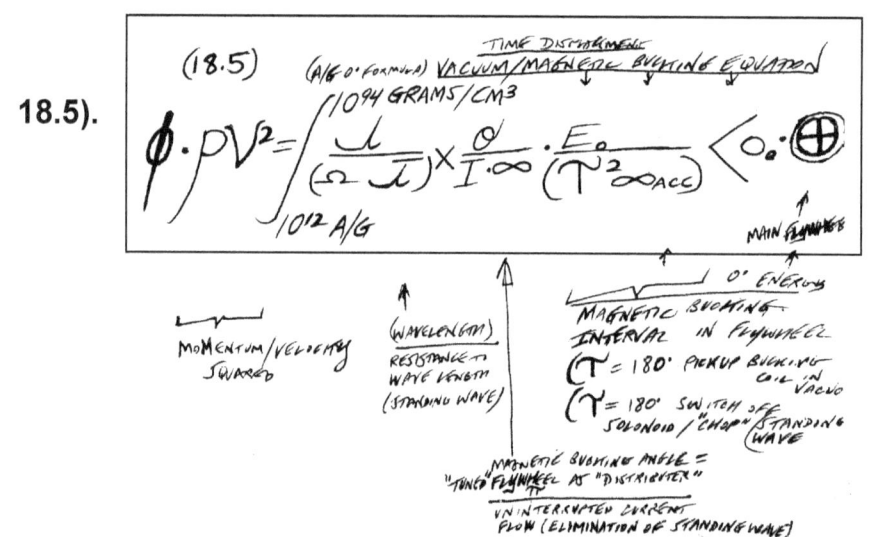

**30).** 
$$\equiv P = \frac{MV^2}{2} \sqrt{\Omega_0 / T_0} \, \Big|_{\int_{-\infty}^{t}}^{\int_0^\infty} \, \frac{\frac{\propto}{\infty}}{136 < |\propto| < 137} \atop (A)$$
(ANTI-GRAVITY/ANTI-MATTER UNIFIED FIELD THEORY)

**22).**
$$\frac{\Omega_0}{T} \equiv 0° \sqrt{\frac{P = MV^2 / 10^{94} GRAMS/CM^3}{10^{12} = A/G = LEVITATION}}$$
(ANTI-GRAVITY ≡ LEVITATION FUNCTION)

**2).**
$$\left( S = \frac{\propto}{\infty} = \theta C \sqrt{\frac{\Omega_0}{0°}} \right)$$
(SIMULTANAEITY EQUATIONS @ AXIS "C")

**28).**
$$\sum \phi \cdot \Delta^{33°}_{180} \equiv \overline{\delta}^4 \equiv \overline{\delta} M_\infty = \frac{\dot{\equiv}}{\equiv} = \frac{-0°}{T} \equiv \left( \triangle_{33°\cdot3\cdot2} \right)$$
(SCALAR WAVE INTEGRAL @ (23) PRIME MOVER TORROID ARRAY) (SCALAR OPEN DELTA / 4 DIMENSIONAL STATE (SCALAR WAVE) E/M MASS — ANNIHILATION OF THERMODYNAMICS @ 0° ENERGY & -0° TEMP / 33° "ROCKING" 0.T." TORROID / x3 = 99° / x2 = 180°)
(4 DIM WHITE BLACK HOLE INTERFACE @ 0°)

**23).**
$$W \rho \theta \times \Pi^\infty \equiv 10^{94} GRAMS/CM^3$$
(FLYWHEEL ACCELERATION FORMULA)

**17).**
$$\infty \equiv \left\{ V = f \cdot \lambda < \frac{\Omega}{\infty} = \frac{V_\infty}{f_\infty} \right\} \infty$$
(LIGHT WARP ACCELERATION FORMULA)

**27.5).**
$$\sum_{\propto 136}^{136} \phi_y \cdot \psi \int^3_{\lambda^4} \int_{0°}^\infty = \psi \delta^4 \cdot \Pi^\infty P = \frac{MV^2}{2} \sqrt{\frac{10^{12} A/G}{10^{94} GRAMS/CM^3}}$$
(QUANTUM FIELD IN FOUR DIMENSIONAL EUCLIDEAN HYPERSPACE / OPEN DELTA PARTICLE @ 4D / TETRAHYDREN PHYSICS)
SUMMATION OF QUANTUM FIELD @ ENTROPHIC THERMODYNAMIC BREAKDOWN POINT × (INTEGRAL FUNCTION) ROOT 4TH DIMENSIONAL INTERFACE @ 0° — ∞ ABS / QUANTUM BETA FIELD   271.B.5.

33). $E^\infty$ (NUCLEAR FUSION) ⟨△⟩ $\phi^\infty_\pi$ [$T^{60°} \times 3 = \Omega^\infty$] (NUCLEAR FUSION [HOP WAVE] THREE PHASE ROTARY CORE)

34). $= \frac{1}{P}dp = -d\Phi; \; \Phi = V - \frac{1}{3}\frac{\Omega^2}{(T^3_\infty)}W^2_{ACC}$;
(UNIFORM MICRO SOLAR ROTATING CONFIGURATION)

35). $\triangle^{33°} = \overline{\Phi} = \Phi(A)$, (SURFACE CONSTANT FOR ISOBARIC – ISOPYCNIC SURFACES, ISOMETRIC ROTATING SYSTEM W/ CONSTANT CENTER OF MASS)

36). $r(a,\theta) = a\left[1 - \sum_{n=1}^{\infty} \epsilon_{2n}(a) P_{2n}(\cos\theta)\right], \equiv$
(UNIFORMLY ROTATING BODY FOR MICRO-SOLAR FUSION EVENT)

37). SIMULTANAETY @ AXIS $\subset$ $\left(S = \frac{\infty}{0} = \theta^C \frac{\sqrt{\Omega}}{0°}\right)$
$\equiv E^\infty$ (PERPETUAL NUCLEAR FUSION)

36). LINE ③⑥ (THE CLAIRAUT-LEGENDRE EXPANSION)

$$\boxed{r(a,\theta) = a\left[1 - \sum_{n=1}^{\infty} \epsilon_{2n}(a) P_{2n}(\cos\theta)\right], \equiv}$$

(UNIFORMLY ROTATING BODY FOR MICRO-SOLAR FUSION EVENT)

$$\boxed{\begin{array}{l}\text{SPACIO TEMPORAL WARP EVENT}\\ \pi \cdot r^2 \; A/G \; (P = \frac{MV^2}{2}) = \pi^\infty (T \cdot r^\infty) \equiv 0° \equiv \\ C = 60° \cdot 3 \triangle = 1.41(C\sqrt{2})\end{array}}$$ QUANTUM LEAP

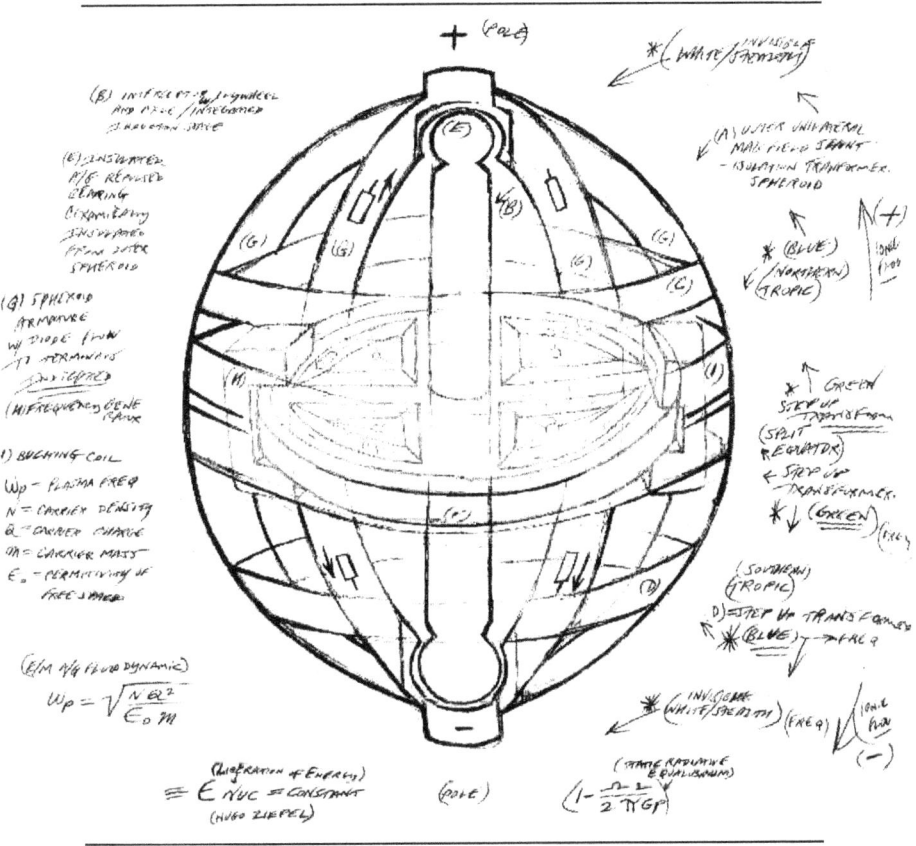

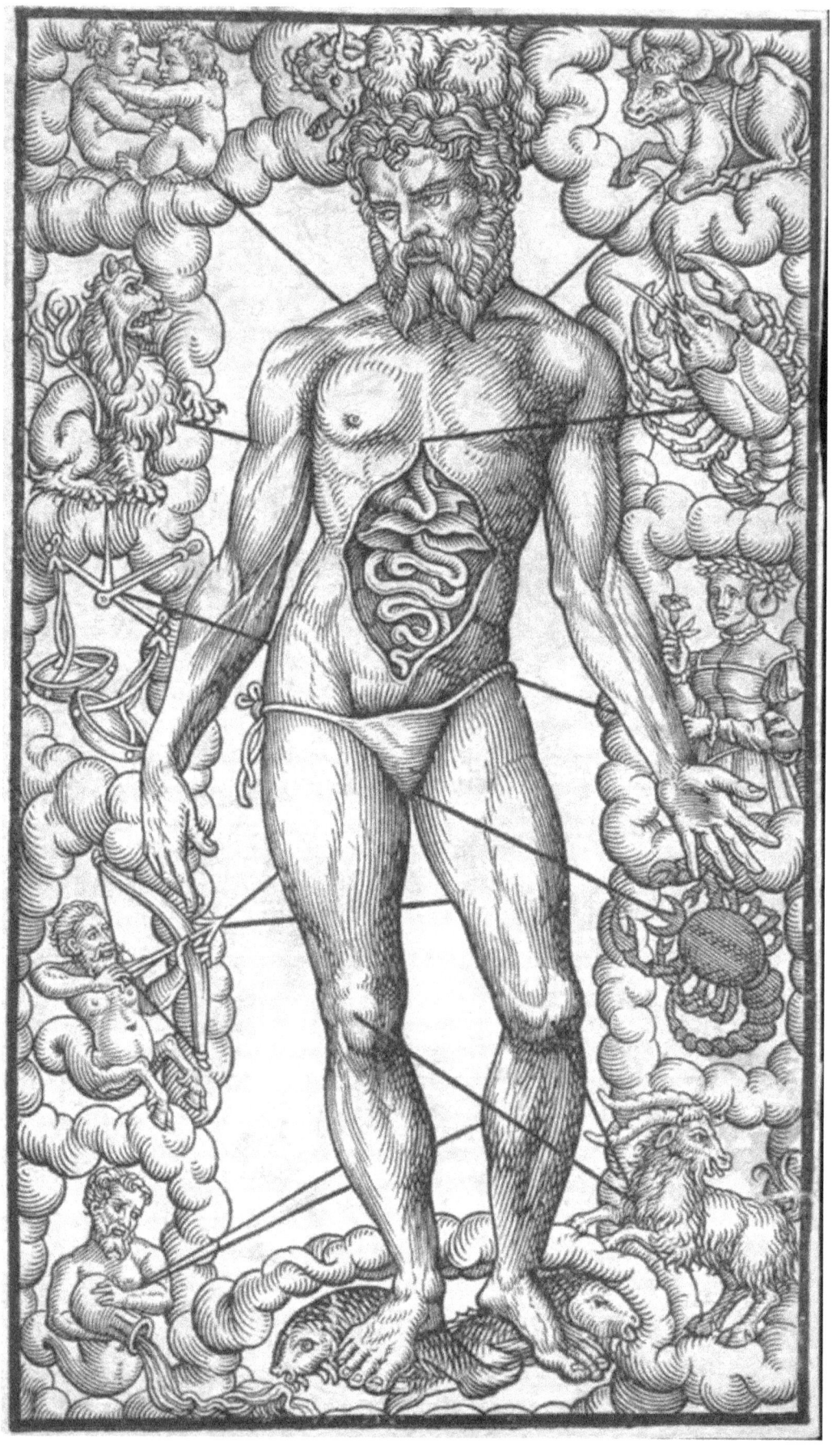

# The Caducean Hieratic Of Matrix Synchronicity

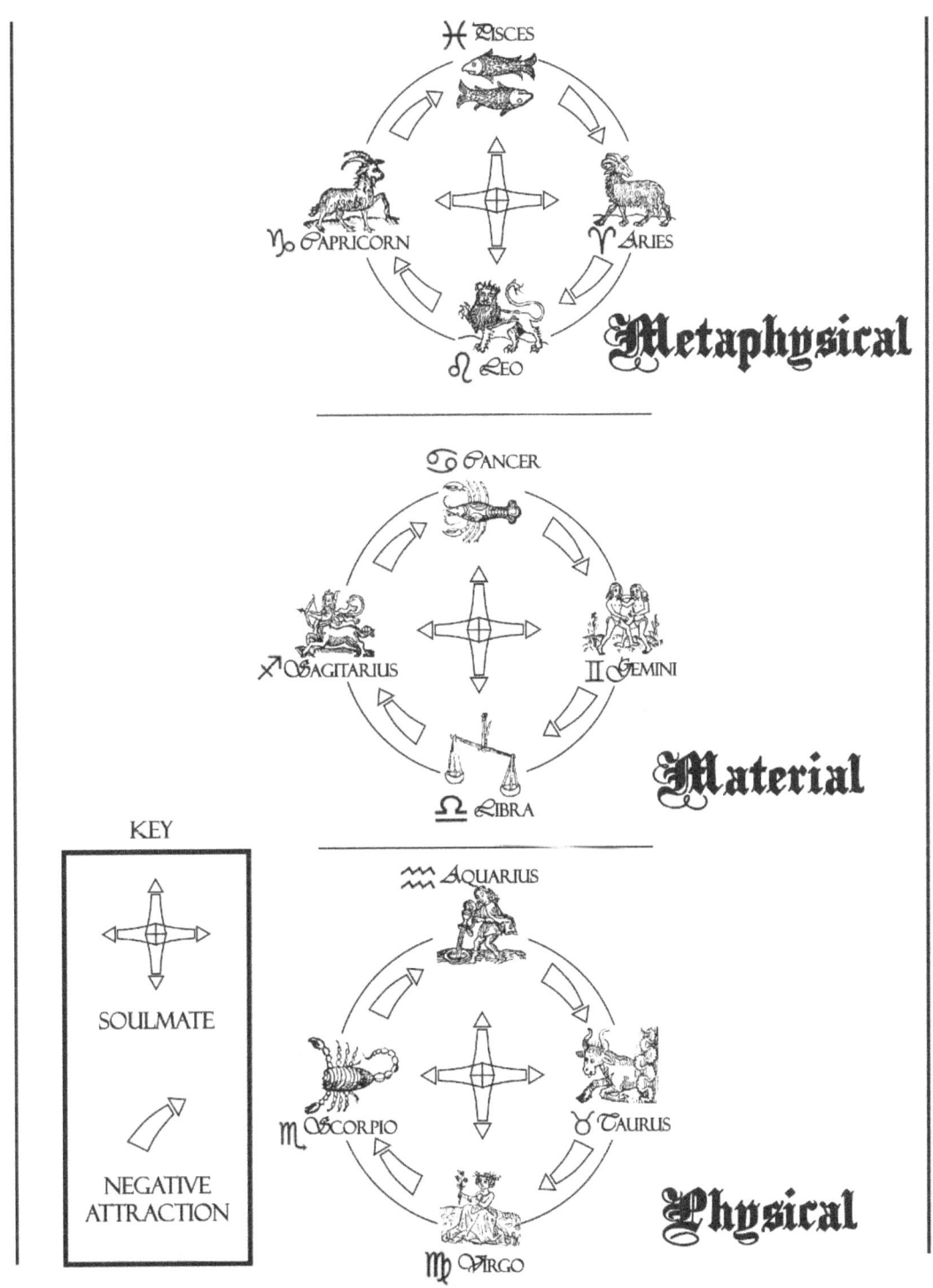

Macrocosmos and the Microcosmos, Scientifically Fathomed, and Philosophically Contemplated, Become Scaled Spectrums, Reflected Universal Cosmogony - So Too, We Made of Stars, Little Gods of Earth, Divinely Correspond, To The Astrophysics, of Our Natal Birth, In Synastry Astrology - We Walk The Earth Alone, Searching For Our Other Half - Synergy, as Fusion Energy, The Primordial Dynamic, That Drives The Universe, Perpetual Engine, Accelerating Faster And Further - Attracting Matter, Bonding Energy, Expanding Space, Periphery Infinite, Unto Itself - Sacred Geometry, Synthetic Attempt, For Geometer and Architect, To Replicate, Such First Form Modulus, To Derive From Mathematically Abstracted, Then Constructed Form, Energy Containing Trajectory, of Ballistic Parabola, To Golden Section, Labyrinthine Upward Spiral - To The Platonic Solids, Endlessly Combine, Creating Vertices Matrix, Inducting Free Energy - Correct Alignment, of Constituent Elements, Accrete The Greatest Valence, In Aesthetic Orchestration, Withstand The Greatest Forces of Nature, and the Ultimate Test of Time, Becoming Iconic Wonders, of the World - With Greatest Precision, They Become Pendant Jewels, Structures That Transcend, Conventional Fantasy - Empirically, They Produce, a Sensorium, of Sacred Space - Transcendent Tranquility, Beyond, Self Witnessing, Mundane Condition - Nothing Is Coincidence, In Such An Ordered Universe, The Closer, The Mimetic Approximation, To The Natural Structures, That Provide Organic Stability, Generating Living Energy - The Greater Complexity and Geometrical Precision, Brought To Innovation, More Perfect It's Dynamic Performance, As Advanced Architecture - Designed as Machine Model, a Biomorphic Replica - Beauty And Strength, Then Assured - Industrial Design, of Any Kind, Accentuates and Attenuates, Natural Capability, Quantum Leaps Further - Anything Short of This, Is Slight of Hand and Charlatanism - Corruption, of Archimedean Laws: Integrating Simple Machine, Into Synergistic Complex Models - Of The Plastic Arts, Architecture, The Most Utilitarian, Must Take Into Account Science, and Practical Value, as Containment Field, For Living Function - Ergonomic Scale, To Human Purpose, In Terms, of Sacred Space, Raising Power/Energy Factor - If The Sacred Geometry Design Succeeds, Like Completed Electronic Circuit, The Flow Of Energy, Will Remain Potentiated and Prevalent - Inducted Environmental Free Energy, Unto Itself, Enhancing Power Factor

Fusion Physics, Is Not Limited to, Twin Tokamaks, Cyclotrons, Linear Accelerators, Magnetic Containment Fields - Is It Not Like Solar Physics Generally, Everywhere, At Once - Where Material, Distance, Scale, Proximity, Contribute to a Cumulative Energetic Effect, a Gestalt Field, A Mesmeric Charisma, All Attempts To Define, It's Invisible, Yet Ubiquitous Force - As Implied In The Alchemical Text, The Marriage Chamber, Purpose Built Architecture, For Soulmate Fusion, The Invisible Castle, of Teresa of Avila, With Solarium, Upper Most Room, The Relation, Between Spiritual Man, And Sacred Geometry, Described Time and Time Again, As A Dynamic Interaction, of Rarified Architected Space, and Ritualized Metaphysical Function - Not A Matter To Be Dismissed, Or Taken Lightly - Through the Better Part of Human History, Architecture Served, This Precise Purpose - It Is Only Since The Age Of Enlightenment, The Industrial Revolution, That Paradigm Shift Has Occurred, To Commercial Function, The Mercantile Become, The New Religion - Subsequently Sacred Geometry Architecture, Has Been Relegated, To New Age Ephemera, Metaphysics - However The Existential

Banal Sterility, Immoral Intent of Corruption, Corporatized Architecture, Will Become It's Own Undoing - The Hive Mind Mentality, of the Promiscuous Crowd, Untenable in the Growing Metaphysical Awareness, Paradigm Shift in Society - The Human Longing, For Purpose and Meaning, Bringing Sacred Geometry Architecture, Back Into Demand, In Future Design Vocabulary: Coherent, Functional, Energetic, Ergonomic, Synergistic

Just as Fusion Physics Advances, Showing Dynamic Energy Reactions, As Spiritualized People, Begin to Bilocate and Find, Their Soulmate Partners, Rarified Space Must Then Be Provided, To Accommodate, The Increased Levels, of Spiritual Power, Becoming Commonplace - Apparent In Indigo Children, The Coming Generation, Will Not Tolerate, The Old Fettered Constraints - The Business as Usual, Inherited Culture, Passed Down To Them, With Ignorant Negligence - I Doubt My Own Story, Will Be Recast, Plagiarism, Thing of the Past - Computerized Accountability, Will Make Such Intellectual Fraud, Difficult, Architects, Designers, Developers and their Lawyers, Their Culpable Antics - Will No Longer Function, In a More Transparent Society - I Suspect the Status, Of the Creative Worker, Will Experience Renaissance - As People Require More Aesthetic, and Intellectual Stimulation, In Their Enhanced Lives - As World Design, Through Computer, Becomes Competitively, More Complex and Sophisticated - As People Begin, To Re-Embrace, The Neo-Classical, Geometrical Constants, The Modern Semiotic Redefined We Are Often Judged, By The Things We Leave Behind, Be It Profound Or Lacking, Before Future Eyes - Where We are Deemed Genius, or Neanderthal, Not Only in Production, of Architectural Work, But The Managerial Design Process: Intellectual Plagiarism, Indicates Many Things, Corrupt Symptoms, That Should Never Come To Pass - Bait and Switch Game, Upon the Public, Detrimental To Highest Culture, Made Thus Low - Shall We, As A Culture, Bare Silent Witness, To Such Dumb Show - Straw Dog, Empty Suits, Masquerading Supermen - Who Is Manning The Watchtower, The Castle Gates, Who Is Minding, Cultural Heritage, Provenance - As These, Malingering And Slothful, Uncreative Dilettantes, Pick The Bones Clean, From Legitimate Creators And Inventors, Like Thieving Highwaymen - Are We To Relive, The Elitist Nightmare, of The Gilded Age, With Robber Barons, Monopolizing Everything - Above The Law, Accusations of Plagiarism, Not Applied To Them - The Public At Large, Pistol Whipped And Intimidated, Too Fearful To Respond, No Care Beyond Survival, Nor Intellectually Grasp, To Fully Understand - I Pray, That We Have Sense Enough and Time, To Course Correct, Our Ethical Sensibilities. The Creative Artist, Must Not Become, Anathema, Endangered Species, For Surely The Rest Of Society Will Follow, Leaving Imbeciles

The Solomonic Key, to the Path of Redemption, Transparency in Philosophical Dissertation, Truth Seeking Dialectic, Over Sophist Diatribe, The Logic of Reality, Triumphing, Over the Illusion, of Temporal Hegemony - Innate Wisdom, Endowed Within All Of Us, Awakened, When The Call To Justice Reverberates - Fabrication of Dissimulation, Will Never Make Us Fools - Verification, of Fact and Law, Ensures Intellectual Survival - Bread and Circuses, Cast Into The Promiscuous Crowd, Baring Blood Sport, of Gladiatorial Spectacle, Rears It's Malevolent Hydra Head, In Times of Corrupting Decadence - Murderous Holidays, To Satisfy the Madding Crowd - Who Bare Oppression, With Affection Smile - Meanwhile In Secret Catacombs, The Remnant Of

Humanity, Whispered Wisdom Speak, Upon Intolerable Debasement, Crumbling Society Furthermore Upon The Frontier, Barbarian Hordes Frenzied, Sensing Internal Weakness, Metastasized Within Crumbling Walls - Whitewashed Graffiti Bleeds Through, With Anonymous Vitriolic, Censuring and Book Burning, Purges Begin - The Voice of Dissent, Forms Chorale of Rage - Disintegrating Illusion Reveals, Stark Reality Discordant Misalignment, Against Nature Widens, As Manner and Custom, Engenders, Transvaluation of Morals - Legitimized Crime, With Oppressors Zeal, Replaces The Republic, With Cesarean Strong Man - The Devil is a Legalist, And Constructs a Plan, Perjury and Plagiarism, Replaces Articles of Confederation - Perpetual War Ensues, Against Protest and Invasion, Culture and Philosophy, Become Weaponized Tools, of The Pundit - Demagoguery Replaces, The Voice of Logic and Reason - Baffling Enigma, Made To Distract and Obfuscate - Reforming, of Founding Principles, Harkening Back, To The Past - Precarious Remedy, That in Anachronism, Cannot Last - Creativity Must Flow Freely, Intellectual Evolution Embraced - Replacing The Proliferation, of Corrupting Lies, With Linguistic Dialectics and Cultural Semiotics, Of Higher Truth - Abandoning The Vacuity, That Has Manifested, in Ignorant Epidemic The Tools Of Ascent, Are Wisdomatic, In Nature - Original Thought, Means of Production, Intellection Property Rights, Must Be Thus Protected: If Mankind, as a Whole, Is To Derive Benefit - Thought By Committee, Unintelligent Design, False Authorship, Are The Terminal Symptoms, of Fall and Decline - Genius Innovation, Intellectual Invention, A Solitary Act, Of The Inspired Individual - Leave Him To His Solitude, Disturb Not, His Vision Quest, Steal Not, His Promethean Fire, To Make Mockery, of the Creative Process - The Plagiarist, is First To Judge The Creator, Discredit Him, Engineer His Martyrdom - How Many, Such Transgressions, Does It Finally Take, To Brain Drain Eviscerate, The Intelligentsia - Reaping Short Term Rewards, By Intellectual Theft - Laying Waste, To The Minds, That Bring Forth New Discovery - The Long Term Juggernaut, Brings Cultural Desert, Unable To Adjust Or Navigate, In Constantly Evolving World - What Message Does It Send, To The Next Generation, Savant Genius, When Rumors Fly, Corporatized Theft - Why Waste Their Time, Put Life In Jeopardy, To Feed The Maw, That Would Consume Them Whole Indentured Slavery, Ostracized, When Noncompliant, To Be Servile and Passive, As They Are Robbed: Brilliance Inherently Finds Way, in Darkness, Creativity, a Power, Transcends, Temporal Condition - I Suspect They Will Find Their Own Way Home, Rather Than Acquiesce, To The Avaricious Caprice, Of Those Who Would Condescend, Limit Them - The Grotesquerie, of Collectivized, Group Think, Unwieldy, In Lowest Common Denominator - Can Only Serve, To Protect It's Interests - Requiring Intellectual Theft, To Vampiric Sustain Itself - My Critique Gleaned, From Direct Experience, Analysis of Events, That Have Transpired: Reading, The Writing on the Wall, Transposed into Book: Like Neberkenezer, Let Them Writhe and Crawl, Upon The Floor, Lycanthropes, in Sheeps Clothing, Geniuses of Disguise - Let Their Stolen Mantle, Fall Off, Let Their True Identity, Be Revealed - When They Cannot Explain, Prophetic Message, Publicly Posted, Going Insane, Attempting Code Breaking, With Interpolation It Grieves Me, To No End, Exertion of Energy, Combating a Wrong, Ingratitude of Highest Order: Steal The Work, Impersonate The Author, Your Life a Fraud, Your Botched Design, Irrelevant and Meaningless: Wind Swaying Phallus, Plagiaristic Vanity

Vilest Cut Throat Piracy, Upon the High Seas, In Disputed Merchant Lanes, Off International Waters, Crews Of Ostracized Criminals, Seeking Refuge, On Floating Bark Of Doom and Anarchy, Anathema Unto, Themselves, And Anyone Who Crosses Paths - The Skull and Crossbones, The Only Flag, for the Nationless, Seaward Scavengers - Strangers, in a Strange Water World - Landfall Awaits, Imprisonment or the Gallows Noose, Condemned Like Flying Dutchmen, To Oceanic Exile - Their Notorious Port, Deserted Islands, Where Their Treasure Horde, Often Rots, Cryptically Mapped, and Left Behind, Forgotten - So Strong The Impulse, To Destroy, Steal, and Amass, in Ravaging Theft, and Sociopath Secrecy - Like Legends, of the Fire Breathing Dragon Lair, Talon Gripping Pearl, Forevermore - Names Associated With Nihilistic Death, They Define, The Lowest Form of Existence, Excluded From Civilization, By Dress, Decorum, Repulsion, and Bad Manner - Hard To Disguise, Such a Motley Crew, They Are Bound Together, By Criminal Code, Blackmail is Their Sworn Allegiance, Communal Hatred, Is Their Social Glue - Hated Upon Sight, Cannon Readied, With Ball and Grape, To Welcome, Their Appearance, With Lethal Repulse - Flotilla of Pandemonium, A Ship Of Fools, Navigating Inevitably, Toward Their Own Doom - Scum of the Earth, Flotsam And Jetsam, of the Seven Seas - Uncharted Waters, Mark Their Watery Graves, Cast Off, Like Detritus, By Their Felonious Mates, Mutiny and Mayhem, Driven Compulsion, That Fires Their Hearts, Soul Sickens, Their Minds - Even The Vessel, They Are Virtual Prisoners Upon, Stolen Token, Trophy of Naval Warfare, Their Vestments, Derived From The Stripping, of Sailors Corpses, Nothing Originally Owned, Polyglot of Plundered Cultures, Romanticized, In Later Days, Curse To Those Who Would Subdue Them: Biblical Plague, Upon The Earth and Sea, Locust Swarm, Of Jettisoned, Lost Humanity What Difference, Does It Make, If Under Corporate Guise, In Raiding Boardrooms, Their Contemporaries, Plot in Cut Throat Secrecy, Function By The Same Code of Honor Among Thieves, As They Divest, The Wealth of Nation, Cities, Individuals, With Impunity And Feckless Guile - Let Us Call This, For What It Is - Plagiarists are Modern Pirates, That Is All - Their Methods Refined, Adapting To The Land, Like a Poisonous Reptile, That Has Learned To Breath, Upon The Shore, Wandering Further Away, From The Polluted Swamp, That Bore Them - Eventually Donning The Trappings of Architect, Developer, Lawyer, With License To Steal - "Architectural Appropriation" the New Terminology, For Intellectual Piracy - Their Designs, Like The Sartorial, Pirate Garb, Stripped From, Sailors Corpses, Wreaks of Cadavers Stench, That No Bleaching Can Deodorize: They Call This Creativity, Innovative Design, When In Point, of Fact, It Is Pirates Cutlass, Vain Butchery - I Would Call Them Thieves, I Would Brand The Liars: However They Would Relish, Such Accolades - Such Men Have No Spine, No Morals Or Scruples, Their Blood Runs Cold, Like Metaphor, Slithering Reptiles, Remorseless As The Scorpion, Who When Treated Humanely, Stings Their Savior, To Repay The Favor: The Proverbial Answer, Always Given: "Well Didn't You Realize, I Was A Scorpion?" - If There Is Any Doubt, As To The Nature, Of These Plagiarists, I Hope This Book, Has Created, an **ORIGINAL** Biological Phylum, For Them - **Reaction Formation**, Against Their Own Profession, Like Pyromaniac Firemen, Or Criminal Policemen, Corrupt Politician, Traitorous Spy - Have Known People Like These, They Are, **ONE OF A KIND -** Like Pirates Crew, Upon Shipping Lanes, **Secret Cabal**, As Thick as Thieves What a Wicked World, these Tyrants Bring, New Babylon, of Stone Effigies, Left Behind

As Fitting Monuments, To Misspoken Blasphemies, Memorializing, Horrendous Deeds, With Stolen Thunder, Foulest Architected - Blame It Upon, The Eternal Night, Within Their Vacuous Souls, and Blackened Hearts - Engorged With Spoils, Derived From Higher Minds, To Inflate Self Worth, Beyond Scale, Value or Reason - Arrogant in Denial, They Live As, Impostors and Frauds, Like Chameleons, They Camouflage Their Audacious Crimes, In Plane Sight, To Glorify Stolen Plunder - Reinforced By Heretical Lies, of False Provenance, That Ring Hollow, In Truth Obscuration - The Propagandist And Demagogue, Repeat and Simplify, Until Many Are Led, Into Damnation - Their False Idol Worship, Cast With Pot Metals, Unsluiced, From Cracked Foundry Crucible, Gold Plated - With Cracking Feet of Clay, Tumbling Downward, In the Temple of Dagon, While Blind Sampson, Inspired by Ultimate, Blind Judgement, With Unearthly Power Razed, Such Philistines Assembled, To Rock Crushing Ground - Pretense, Arrogance And Pilfery, Megalomaniac Gravity, Makes Such Effigies, of Wicked Men Fall - Hate Stricken, From The Book of Life - Their Monuments Falsity, Destroyed, in Pregnant Time Intolerable Grim Reminders, Pariah Memory, Loathsome Footnote, in Human History, Glaring Examples, of What Not To Be: Taboo Objects, of Universal Rebuke and Scorn - Impostor Existence, Reduced to Pathetic Farce - The Plagiarist, His Own, Greatest Masterpiece, Erroneous Falsehood, Copied Human Being - Phantom, A Shadow, Once Creative Artist - Mutated into Antipode, of Counterfeiting Thief - Release The Kraken, From the Oceanic, Cave Floor - Beckon, The Leviathan, Moving Atoll, Upon Sounding Main - Draw Forth, Charybdis Whirlpool, To Purge, With Cyclonic Waves, Of Piratic Sin - Summon Forth, The Chimera, To Pluck Upon, The Remnants, Washed Ashore, Sea Piracy, Released Dashing, To The Sirens Calling, Upon Jagged Coastal Shoals, Monstrous Sport Targeting - Exasperated, As I am, To Be Tread Upon, By Lesser Men - My Warriors Blood, Brought To Horrendous Boil, By Horanguing, Pigmean Races - Like Gulliver, in His Fantastic Travels, Rebuking Lilliputians, For Minutiae Pettiness - Satirical, As This May Sound, It Is Unfortunately, The Case - Like Gollum Colossus, Animated Guardian, With Sealed Phylactery, of the LOGOS Tetragrammaton, Protecting Sacred Community, From Further, Invasion Trespass, Unorthodox Disgrace, Corrupting Outside Influence, Contemptuous Chronicled Abuse To Judge Aright, With Discerned Reckoning, Degree of Trespass, Meting Out In Divine Retribution: Such Dispensation, Poetic Justice, Allotted By The Gods, Beyond Mortal Power, They Intervene, To Bring Such Justice About - When Things Have Gone, Too Far, Mankind Can No Longer Regulates Itself - The Elemental, Primordial Powers, Are Released From Inaccessible Vaults, To Wreak Havoc, With Unsuspecting Damnation, Upon Unredeemable World - Targeting Those, Evil Masterminds, Who Would Make It So - Marked For Certain Death, Special Place In Hell Reserved - For Making All Things Around Them, Wayward - Eviscerating, Their Malevolent Memory, and Undue Influence Purgation Of Their Entire Genetic Line, Only The Wrath of God, Can See This Through: Given Ample Opportunity, To Repent Their Ways - They See This Only, As Further License, Laying Down, Continuous Abuse, Upon Others, Mistaken By False Notion, They Are Impervious - Flaunting The Law, Continuing Onward, On Descending Course, Marveling, At The Cruelty And Larceny, They Have Perpetrated - Mistaking Criminality, Foul Enterprise, As Intellectual Genius - Thus Raised Up, To Audacious Height, They Will Be Stricken Down, To Achieve Greatest Distance, Steepest Plummeted Falling:

Stripped Of Everything, But Pain And Suffering, Misanthropic Exile, Will Soon Be Waiting, Social Leper Banishment, Eschewed, Rebuked and Pariah Shunned, From Every Righteous Doorway - Stones Cast, Upon Them, To Drive Them Thither and Hither Hence, Like Desert Eremite, Cave Dwelling Hermit, Bridge Harboring Troll, They Must Suffer, Their Own Repugnant Company, Vilest Reclusion - No Meditation, or Prayer, To Assuage, Their Thoughtless Guilt - Unexamined Life, Not Worth Living - Stripped Of Everything, They Value Most: False Title, Artificial Status, Soporific Peer Respect, Scandalous Fraternal Association - Once The Rotten Fruits, Of The False Labor, Have Been Discovered, As Stolen - Their Own Orchard, Barren, Fruitless, Weed Overgrown - Embarrassment, To The Company, Once Dubiously Kept, Reputations Besmirched, **PLAGIARIST** Guilt, By Association, The Keys To The Kingdom, Taken From Them, No More **Cloud Castles**, To Design, Wind Wavering, in Autumnal Mist - Towers of Babylon, They Would Recreate, Abominations of Desolation: **THEY MUST STEAL TO CREATE** - Sad Commentary, On The Declining State, of Decadent Things - How Many Authentic Architects and Designers, Were Crushed, To Make Room, For These, Rank Impostors, It Boggles The Mind, To Consider This - Where Were The Many, Who Were Called Upon, Only To Be Exploited, Their Original Concepts, Purloined, **BUTCHERED AND PLAGIARIZED**: May This Book, Shed Light, Upon This, Growing Phenomena - Hackneyed, Ham Actors, Billowing Bombastic Lies, Felonious Raconteurs, Upon an Imaginary Stage, Surreal Reality, Becomes Schizophrenia, As Their False Act, Wears Tiresome, Redundant, in Tedious Repetition: Waxing Thin, In The Telling, Devoid of Convincing Detail, Emotionless, Contrary Currency - The Staging, Of Stolen Props, Festooned, With Well Placed Sycophants, No Longer Offer Chorus, Bloated Toadish Affirmations - The Narcissistic Vestments, Meant To Self Delude The Wanting Thespian, Only Serve To Unravel, The Mummified Corpse: Reliquary of Doubt, Thing Of Filth, Misplaced Archeology, Historical Provenance **LACKING**, Contextual Intellectual Irrelevance, With Forked Tongue Removed: Unable To Pander and Dissimulate, Like Gutter Snipe - Painting In Detail, This Character Portrait, Modern Phenomena, Of Plagiaristic Piracy - **Caveat Emptor**, "Let The Buyer Beware", Should Be The "Professional Shingle", Forced To Wear, Around Their Neck, But That Would Be, "Truth Telling": Incapable Utterance, Of Still Born Thief - From Cradle To The Grave, Such Spawn, As These, Dissimilate, Until Their Wagging Tongues, Rot Right Out Of Their Heads - Such Welcome Silence, Creators Catharsis, To Hear No More, Subtle Fabrications, Perjured Testimonials, Of Stolen Creation - Enough To Have, My Original Design Mangled, Into Titanic Skyline Eyesore: Creative Frustration, Is Over Now, Having Completed, My Original Task, Memorialized, My Warranted Complaint, Set Into Published Form, Free Unexpurgated Literature, Chastised Those, Who Saw Fit, To **GREATLY** Wrong Me, To My Hearts Content: **Brightbill, Bergen, Papert, Childs, Bloomberg, Doctoroff, Calatrava, Piano, Silverstein,** You Are Beneath, My Lowest Contempt - Thank You, For the Dubious Honor, **Architectural Fetishism,** Appropriating My Design, Destruction of My Business, With Plagiaristic, **"Slight of Hand"** - May The Actual Truth, Of Design Origin - B(I)oomerang, Right Back, Into Your Facetious Faces: "Three Men Can Keep A Secret, When Two Are Dead" - **Benjamin Franklin** - If Franklin, Were Betting Man, Would Lay the Odds, of **MAJOR SCANDAL @ 9 to1** - Inquisitive Investigation, Find Weakest Link, In The Chain of Custody: **"Impersonator"**

Archimedean Leverage, The Prying Axis, "Reality Point", Upon Scandalous Avalanche, No Moss Will Grow, Upon Rolling Stone - Tipping Point Iceberg, Conceals Glacial Mountain Below, Truth Concealed, Will Ship Wreck, The Soporific Navigator, With Hull Breached Overflow - Newtonian Joules, of Herculean, Perpendicular, Torque Quotient, Thermodynamic, Moment of Shear, Upon The Proverbial, Rusted Bolt - Parabolic Swinging Arch, Train Maul, Kinetic Travel, Upon Spiking Head, Crucifies The Criminal The Fulcrum Power Factor, Of Levered Instability, Where Controlled Collapse, Engineers, Imminent Momentum - "Archimedean Claw", Capsizing Pirate Ships, Off Walled Battlements, of Syracuse - Triform Derricks, Rampart Arrayed, With Hooked Chain-Fall, Block and Tackle, To Employ Deadly Physics, Upon Floating Invasion Plague - Siege Engine Armada, Flipped Over, Into The Sea, Like Promiscuous Shark Chum: Superior Minds, Meet Conflict Head On, The Idea, Remains Always, On Their Side - Overturning What Was, By Invention, Thought, and Acumen, Attacking - They Will Contemplate, With Precise Calculus, Perfect Strategy, Lay The Enemy To Waste - With Cold Mechanics Of Reason, They Take No Prisoners, No Quarter Given - They Are Aware, No Sympathy, Must Be Given The Wicked, Having Been The Primary Targets, Of Their Chronicled Derision - Like Archimedean Lever Point, The Tables Inevitably Turn, The Thing, They Sought, To Covet Most, Becomes The Secret Weapon, They Most Abhor - The Mind From Whence, The Creation Came, Will Protect Itself, From Further Onslaught - How Can You Outmaneuver, Inventive Genius, Psychic Clairvoyant, Righteous Prophet, Inspired As They Are, As Actualized Individuals - Against Mock Centipede, Of Lock Stepped, Dilettantes - Proffer The Question To You, If You Would Trespass Against Them, Beware,  For They Are The Children of God, Angelology Protected - They Are Guarded and Avenged, With Metaphysical Fury, Incalculable - Their Words Speak Fire, Their Reckoning Voice, Human Thunder, Reverberating, Throughout The Ages, Becoming Part of Eternal History - No Amount Of Lying, And Propagandist Obfuscation, Can Dull Or Hide, Their Original Brilliance, Deny Their Creative Progress, Shining Forth, Like Noon Sun - Like Phantom Pain, The People Question, Where Such Incredible Things, Originated - Many Have Laid False Claim, Egotistical Plagiarism, Paid The Dearest Price, Dismissal, And Disgrace - It Is Human Nature, To Despise Such Men, They Steal From All, When The Halt The Futures Progress - Attaching Notorious Name To It, Like Cheap Gilding, Upon Pagan Shrine, Abomination Of Desolation, Idolatry Devoid of Sense and Purpose, Uncontemplated Thing: Purloined Bauble, Facsimile of Greatness, Reduced to Statuette Figurine, Miniature Curiosity, That Invokes Derision, Haunted By The Original Thing, Copied, That Casts Giant Shadow Upon It - General Populace, Not Easily Fooled, They See Cracks And Imperfections, Of The Butchers Hand - They Look And Ponder, And See The Spector, of Unfinished Thing: Periphrasis, of Lost Philosophy, Fragment of Masterpiece They Wonder, **HOW,** Did That Fragment Get There, Like a "**SHARD**" from Museum Artifact, Blended Into, Retaining Wall, Guarded by **JUNK YARD DOG** - Puzzlement, Not Assuaged, By The "Mosaic Artist" Who Placed It There - Calling, **"Ode To The Grecian Urn"** Salvage, Broken Along The Way, Mortar Plugged, Into Camouflage The Ravenous Beast, Like Saturnine Cerberus, Protects Lie, Of Hades Portal - Soon Herculean, Twelve Labors, Brought To Bare, Will Tame, The Ferocious Monster, Retrieving Mystery, **MISSING ARTIFACT**, Culprits Judged and Blamed, Lost Masterpiece Returned

# Penitentiary vs Community Habitation

The Future is Yours, You Can Do Anything, So Sayeth the Samurai Warrior, Turned Anarchic Ronin, Rendered Masterless, With the Breakdown of Japanese Feudalism, in The Existential Quandry, of Their Obsoleted Purpose, Become Hired Mercenaries, Wandering Men at Arms, No Castles To Protect and Guard, Need Arose To Reinvent Themselves - Soon Persecuted as a Growing Danger, to Milder Civilization, The Invention of the Gun, Outmoding Samurai Sword Katana - Enemies of the State, Forced To go Underground, Forming Secret Society of the Ninja - Musashi, Cave Dwelling, on Kishu Island, Poet Sculptor, Greatest Example of the Samurai, Unbeaten Warrior, in 60 Duels, by the Age of Thirty - He Chose Creative Seclusion, After His Greatest Battle, Carving a Deadly Club, From a Rowing Boat Oar, Time His Shore Landing, With The Rising of the Sun, at Dawn, Waiting For The Solar Glare To Hit, His Opponent's Eyes, Braining Him On The Head, With A Single Blow - As His Entire Warrior Clan, Look On With Horror and Disbelief - As Musashi, Disembarked, With Trusted "Second" - Concluding His Demonstration, of Perfect Strategy, "The Five Rings", a Written Testimonial, to the Art of War - A Strategic Philosophy, Become So Powerful, When Men Are Taught, They Can Do Anything - Yakuza Koreans, Were Brought In To Stop Their Espionage Rampage, Later Forming Criminal Syndicates, With Their Corporate Sponsors, Attempting To Keep The Samurai Secret Brotherhood, at Liberal Distance Stalemate Achieved on a Prison/Penitentiary Island - Purgatory, of Power and Control

Jeremy Bentham's Mummified Head, Autopsy Decapitated, Then Laid to Rest, in the Board Room Cabinet, of the London School of Economics, Presiding Cadaver, at Every Meeting, Specified in The Philosopher's, Last Will and Testament - Father of the Penitentiary Architecture, Sacred Geometry Inverted, To Create Penitent Cells, The Miracle Cure, of Enlightened Penology, Recreating the Monkish Cell, of Cloistered Monastery, Prison Yard Arrayed With Guard Towers, Unscalable Walls, To Segregate And Ghettoize, Incorrigible Criminal Populations - The Way of the Dungeon, and Prison Chamber, Abandoned For This Philosophical Method, of Civilized "Humane" Treatment, Time and Reflection, Upon Ones Perpetrated Crime, Considered Punishment Enough, To Cure The Criminal, Within Hermetically Sealed, Architectural Environment - The Bentham Experiment, Now Become, The Penitentiary Industrial Complex, Privatization, Of Originally State Controlled Function, Longer Sentences, Virtual Slave Labor, Factories Introduced Within Prison Complex, Profit Driven, Into a Prison Mercantile I Wonder How Bentham Would Consider This, He Had No Comment, At The Last LSE Boar Meeting - When They Opened The Cabinet Doors, The Oracle Would Not Speak - Jeremy Bentham's Head, Now Penitentiary, Unto Itself - Rotting Away in Solitary Confinement - Perhaps if Bentham, had Embraced Other Philosophies, "The Future Is Yours, You Can Do Anything" - Could Have Been Conceptualized, Architecturally -

Utopian Community, No Walls Required, Voluntary Constellation of United Individual, To Come and Go, As They Well Please - Architectural Design and Urban Planning,

Determines Retention Rates of Participants, Alone - Recourse to Ostracision in the Extreme, Meted Out, as in Direct Democracy, To Those Who Would Endanger, The Freedom of the Polis of Individuals - Not the Other Way Around - Would Be Tyrants, Traitors, Spies, Ever Present Threats in Free Society, are Not Fit, Nor Inclined To Form Utopian Community - Their Tremendous Egos, and Incivility, Are Not Inclined To This - Let Them Join The Barbarians, on the Periphery, Often The Rational For The Ostracized Some Are Better Suited, To Chronic Adversity, Would Rather War Upon Each Other, Than Engender Communal Peace -  To Each To His Own, In The Final Analysis - Men Are Inclined, To Seek Their Own Kind - In The Grand Scheme, Actuarial Algorithm, Nature Of Things - I Say, Let Them Go Their Way, Begone - However if the Epidemic Problem, is the Design Modulus Itself, Crumbling Civilization, Becomes Both The Penitentiary, and the Arch Criminal - The Philosophical, The Creative, The Inventive, Will Flock In Droves, To Immigrate From It - Empirical Proof of Failed Experiment, Nihilistic Ideal, Where The Corrupted Criminals, Now Run The Penitentiary, Thinly Disguised, as Urban Metropolis - The New Architecture Often Reflects This, Variation on A Theme, of Bentham's Prison Monastery - More Akin To Insect Hive, Than Proper Habitation, For Human Beings - Breeds The Psychology, of Form Meets Function, Human Race Reduced To Insect Swarm - Ergonomic Scale, Demography, Architectural Density, In Relation To Natural Environment - The Imposition of the Man Made Synthetic Modulus, Upon Environmental Nature - A Discreet Relationship, That Must Be, Experienced and Examined - Compounding Human Suffering, Social Ills, Or Elegantly Solving Them, With a Philosophical Foresight - Economic Driving Points, When Unregulated, Invariable Serve No One, In The End - Planning Abominations of Desolation, Living In The Maw of the Beast - An Obvious Misconception, of Human Habitation, Misconstrued with Penitentiary Incarceration - The Prisoner as Resident, Pays DEARLY For - Blighted Community, Diminished Life Energy, Imprisoned Minds, To Adapt To Hive Architectural Modalities - Freedom Sacrificed for Luxuriant Security, Beautified, Aesthetic Versions, of Bentham's Guard Towers - Everyone a Criminal

The American Suburb, Yet Another, Prison Encampment - Engineered Social Mortification, From Inception - Designed, To Pacify, Returning Military Troops, Forming Urban Gangs, Post WWII - Luring Them Out Of The Urban City Centers, With GI Loans To Purgatory, of a 100 x 25, Land Allotment, Suburban Single Family Home "Unit", The EXACT FOOTPRINT DIMENSIONS, to Create Perfect Social Dysfunction, and Intellectual Isolation - No Community Formation Possible, in Planned Dysplasia, of Mortified Demographics - The Antithesis, of Clan, or Tribe, Migrational or Stationary, Creating Autonomous Zones of Privacy, Within External Matrix, of Natural Gathering Point, Discreet Intervals - Seen Also, In Village Towns of Europe, Pedestrian Scaled, Self Sufficient, Centralized Locus, of Concentrated Social Activity - Continuously Inhabited, for a Thousand Years or more, Human and Architectural Scale, Harmonious With each Other, And With Surrounding Environment - Adjusted Scale of Dynamic Ergonomics, Architecturally Conceived, Through Empirical Observation, Communal Function, From Indigenous Permanent Materials, Meant To Last For Vast Amounts of Time - Culturally Connected, To Place and Time, Aligned With What Has Been, and What Will Be - Intelligent Design, of the Highest Magnitude - Architectural Hermetic

Interconnection, in Dynamic Matrix, Creates Utopian Community, in Polis Habitation - These Components of Holistic Inter-Relationship, can be applied to Modern Architecture Like Units of Ergonomic Scale, Producing Complex Radiating Systems - Providing The Blueprint for Metropolis, Comprised of Scaled Down, Ergonomic Units, of Viable Community Habitat - The Unsavory Alternative, of Urban Sprawl, Ghettoization of the Human Spirit, Profoundly Effects, our Individual Potential - "The Future Is Yours, You Can Do Anything" - A Clarion Cry of Futurity, In The Existential Modern Wilderness -

The Pando Aspen Grove, The Trembling Giants, of Alpine Plateau, The Largest, Oldest, Heaviest Life Organism, Existing in the World - Inter-Connected, by Miles of Root Systems, Giant Trees Arranged, in Arrayed Proximity, Like Sacred Grove - For Optimal Conditions, For Growth, Colony Protection, Expansion - The Most Successful Arboreal Species in the World - Calculation and Investigation, of This Genetically Identical Horde, Concludes, It Exists As A Single Bio-Organism - The Ergonomic and Demographic Geography, of This Vast Forest, Surely Resonates in Scale, With Other Spontaneous Gathering Migration Points, and Stationary Village Communities, With Similar Root System Nexus, Connecting Community, By Discreet Physical Proxemics - Such Synergistic Human Energy, Enhanced, Not Hindered, By Surrounding Architecture - This Defines The Antonym Difference, Penitentiary and Habitation, Between Freedom And Incarceration - Semiotic Relationship of Form to Human Scale, Aesthetic Ergonomics, to Demographic Balance Point, Homeostasis Between, Public and Private Life, Autonomous Creative Zone, To Reflect and Create - Necessary Machinery, That Must Be Accommodated, if an Community Habitation Architecture, vs, Penitentiary Confinement Modality - Encapsulating a Greater Definition, and More Civilized Conception, of Architecturally Designed, Urban Demographic Space - Ultimately Conceived Around The Basic Need and Want For Shelter, And The Universal Human Phobia, of Involuntary Confinement - Call These Governing Factors, of Architecturally Enhanced, or Confined Horrific Experience, What You Will - The Lines Have Blurred Substantially, Somewhere Along The Way - No Human Being, Unless Insane, Would Choose, The Restrictive Confines, Of Incarcerating Cage - Human Population, Not Meant, To Be Stockyard Corralled, Into A Madding Crowd, An Insect Horde, In Directionless Swarming Frenzy - Like Grasshoppers, That Congregate, Like Demons of The Air, Brushing Bustling Wings, Transformed Into Locusts, Horned Protuberances, Erupting Everywhere - Rendering Everything In Their Path, To Desolation - So Too, Human Beings, When Population Stressed, Will Cease To Breed, Become Decadent, Polymorphous Perverse, In Final Stages, They Will Be Driven to War Cannibalize, All Determined, by Gradients of Demographic Compression, Dynamic Ergonomic Scale

Human Instinct, Often Subjugated by Synthetic Intelligence, Errant Philosophy - What Seems Logical On The Surface, Can Be Our Greatest Undoing - Confabulations and Constructs, out of True, With Reality - Will Beget Unnatural Phenomena, Let The Epidemiology of Plague - Modern Economies, Often Hinge on Instability, Planned Obsolescence, Propagandized Market Forces - Conspicuous Consumption, Reduction Of Living Space, Individual Now Rogue From Nuclear Family, The Greatest Human Waste, Per Capita, To Derive Greatest Profit, When Will Such Insanity End, One Begs

The Question, "The Future Is Yours, You Can Do Anything" - What Is To Define, A Modern Vernacular, Vocabulary, Syntax, of Modernity in Architecture, Given What We Know Of Things, That Effect The Human Condition, Both Positively and Adversely - Towers of Babel, Warehousing Mentacides, Incarceration Triumphing Over Habitation, The Enclave and Lair of Beasts - Or Re-Evaluation, to a Human Scale of Being, Contextualized, With the Forethought, of Dynamic Ergonomic Dwelling Place - Sense of Home Abating Wanderlust, Transience of Population, Restlessness Halted Predatory Instinct Grown Dormant Again, Evolution Progressing, In The Manifested Form, of High Civilization - A Tribute, Not a Negation, Of Individual Worth, Humanity in The Human Being, Essential, In Thriving Community - The Agora, The Lyceum, the Acropolis Mount, All Jeweled Fractals, of a Perfect, Architectural Organization, Paradigm - Serving as Best, Historical Congregation Nexus Points, For Philosophical Thought, Dialectical Speculation - The Powerhouse Dissertation, By Which all Subsequent Creativity Derives - The Mind And Heart, of the Utopian Polis, Where Greatest Conceptions of the Ideal, Spontaneously Arise - God Lives in the Machinery, Of Such Sacred Space, Geometers Computed, From Egyptian Mystery School - Pythagoras, Euclid, Archimedes, With Chain, Rule and Square, Derives The Ergonomics Golden Mean, of Human Calculus - Such Intent Was Lost In Later Days, As Human Interest, Turned To Mundane Things - Original Thought, Deemed Non-Material, Became Secondary, To Be Plundered, Plagiarized, Absconded, In A Thousand Syncretic Forms - Bereft of Context and Meaning, Like Architectural Cemetary - The Inter The Ancient Remains, Reliquaries Abandoned, Vital Processes Of Intellection, Free To Speak, No More - On Unhallowed Ground, The Mutated Towers Arisen, Crushing, The Contained Life Force Down, Like Vampiric Gravity - Over Pentagram Points, Satanic Monuments, to Subterranean Baphomet Draconian - Where Are The Dragon Slayers, of The Future, Who Will Build Anew, In Unexpurgated, Unplagiarized, Creative Freedom - I Would See That Day, Come Sooner Than Later - Before The Poem Fire, Is Finally Extinguished, And We Pay No Homage, To The Righteous Ancestors - They Who Left Us Their Knowledge, In Testaments of Stone, Built Sacred Architecture For Us, As Well As For Themselves, Meant To Endure Eternity, To Remember Them, Within Ourselves - Narcissistic Vanity, Has Taken It's Place, Vainglorious Empire, From Dust To Dust - Signifying Nothing, Beyond Material Greed And Lust - A Mundane World of Pathetic Men - Close Is The Time, When Such Intellectual Dwarves, Educated Imbeciles, Autocratic Dilettantes, Will Meet Their Match, In Aeon Paradigm Shift - Vacuous Consumption, Empty Suits, Unequipped For the Technological Quantum Leap, That Is Overdue - Retreating to the Skyline Penitentiaries They Made, Waiting For the Final Lights To Go Out - Genius Eludes Such Demagogues, They Cling To What Conserves and Concentrates, Their Temporal Power - Instinctually Fearing, The Original Idea, The New Philosophy, and the Point of Shear, From the Old Ways, Unto the New Ways of Being - They Would Rather Have Us Worship Them, With Arcane Cesar Toga, and Laurel Wreath Crown, Not Realizing, They Are In Fact, Troops Of Clowns, Mocking Whispers Follow, Them To The Grave - They No Longer Amuse, With Pompous Show, Magic Tricks of Stolen Ideas, Equalitarian Spirit, Meritocracy, Renders the Buffoonery of Megalomanic's, A Universal Thing, of Derision and Scorn Energy Supplied To Feed The Hoax, From Oppressed Individual Expression, Soon

Returned, When the Wellspring of Renaissance, Quickens, Genius of the Philosopher King, Returned to Natural Order, in Utopian Republic - Architecture Plays It's Role In This - Communal Habitation, or Penitentiary Incarceration, A Design Modulus, Well Extending, Beyond Jeremy Bentham's, Encroaching Walls - The Opposing Force of Community Building, A Mighty Bulk Work, Against a Prison Planet - Without The Philosophical Spark of Brilliance, It Shrivels on the Vine, Like Aborted Ideal - The Modus Operandi, For Such Dialectic Ascent, The "Talking Stick" of Algonquin Longhouse, Parliamentary Procedure, Seat of Government, Encapsulated by Long Contemplated Architecture, Housed Generator, of Human Liberation - Reduced at Corrupting Points in Time, To "Punch and Judy" Puppet Show - Constant Reformation, Ogmios-Dialectical Refinement, Skilled Diplomacy, Dedication to Universal Truth, Must Serve, as the Foundation Stones, and Supporting Pillars, of Utopian Constitution - All Things Are Vanity, Beyond This Precept, They Will Not Withstand, The Empirical Analysis, of Philosophical Discourse, Deemed Unrealistic Things, Repugnant Themes, of Sophistry

Once Grandest Discipline, of Plastiform Arts, Subsuming All Other, The Nine Creative Graces, Of Parnassus Muses, Within Architected Structure, All Meld Into One, "Opera" - Required Repository, Creative Milieu, Housing Performance and Display, of All Other Artistic, Linguistic and Musical, Creative Expressions - Rarified Environed, Stage Design, Intended For It - How Debased, Has It Thus, Truly Become, How Fallen, From Olympian Heights, Into Cultural Bankruptcy, Professional Pawn, of Venal Mercantile, Subservient, To Age of Wickedness: Centralized Sickness, Echoed in the Other, Once Sacred Performing, Literary, Musical Arts - Cascading Downward, With Geometrically Increased Velocity - To Irrigate a Cultural Desert, With Running Downstream Sewer Water - Anyone Who Denies This Niagara Cascade, Is A Blinded Fool, In Need Of Telescopic Glasses, and an Awakened Sense of Smell - Mistaking, Pons Eternum, For A Septic Dead River: Brought Thus So Low, In Whirlpool Charybdis - They Cling To Plagiarism, Intellectual Theft, Last Resort, As Stolen Life Preserver - It Is, At This Crisis Point, Paradigm Shift, To Stop The Final Falling, Into The Abyss - Where The Sins, of the Fathers, Through Active Contrition, Are Flaw Forgiven Unto Us, Conceived Utopian Models, Of Humanistic, Philosophical Modernity, Then Arise - By Miracle of Grace, Intuited Talent, Genius Endowed, With Power to Bring Forth Change, The Outmoded Modality, We Struggle With, To Live And Survive - Become Righteously Meant, To Flourish and Multiply - Greatest Battle, of the Adversary, To See Us Turned, To Swine, By Circe, Witch Hexing Grimore, Left With Dim Brute Memory, Of What We Were, Adding Insult To Injury, A Collusion of Damnation - When We Chose, To Abandon Our Tools of Ascent, Our Instruments of Cosmological Navigation, To Gauge, Fathom, Measure, and Self Calibrate, Artifacts Left To Rust - No Longer Meant To Ponder, What We Are, And How To Live Philosophically, We Are Rendered No Better, Than The Most Unclean, of Uncloven Hoofed Animals, and Shell Fish, Bottom Sea Feeding, Detritus Parasites - From Sacred Mountain, Sanctuary Cavern, Above Water Line, of Atlantean Diluvian Flood, Weathering, The Ice Age Onset, Protected Us, From Most Extreme, Primordial Elements, World Change Unleashed - Replicating In Ancient Neolithic Temple Compounds, The Mountain Landscape, Cavernous Mother Uterus - From Such Living Rock, We Received, Earliest Inspiration of Religion, From Those Sacred Spaces,

We Returned, To Receive Prophetic Epiphany, in Latter Day - The Most Ancient People, Are The Children, of the Highland Mountains, Perfected Utopian Cultures, Complex Linguistic Systems, Yearly, Death Defying, Games of Chance, Unbroken For, Twenty Five Thousand Years, Stable Population Growth, Unilateral Self Sufficiency, Cradle to Grave, Community Support Network, Longevity and Mental Health, Shaman Parameter Optimized - Contrary to Malthusian, Actuarial, Demographic Load Tables, Like The "Trembling Giant", **Pando Aspen**, Tree System Grove, Largest, Oldest, Heaviest, Single Biological Organism, In The World - Root System Connected, Over Vast Tracts, Of Mountainous Plateau, Unilateral Undulation, Causing Geological, **Seismic Activity** - Homogenous Societies, Genetically Related, Stand The Greatest Chance, Of Long Term Cultural Survival - A Human Equivalent, Root System Entwined, A Viable Social Units, of Organic Interwoven Community, Facts Flying in the Academic Face, Of Apologist, Soft Science Sociology - Another Perpetrated Fraud, To Camouflage, The Blame Game, Victimize The Victim, Explain Away, The Psychosis Breeding, Anathema, Of "Unplanned Blundering, Ghettoization, By Socially Mortifying, Non-Design Paradigm", Purgatorial Urban Sprawl - Penitentiary Inmate, Within Culturally Mortified, Desolated Landscape, vs, Rooted Member, in Community Habitation, **"Vibrationally Connected"** The Ultimate Choice Is Ours, Remember: "The Future Is Yours, You Can Do Anything" - Ignorance of Fact, Unperceived Reality, Oblivious to the Laws of Nature, Will Be Humbled, And Hovelled By It, In The Length of Days - Biological Structured Phased, And Sequenced, The Machinery of Life, A Perpetual Engine - Stress Placed Upon The System, in Multifarious Forms, Are the Nihilistic Disease Model, of Manipulated Habitation, Anathema To Architecture, Whether Anatomical, or Environmental, A Superturnkey Engineered Universe, Where Both Biomorphic Systems Embrace: One Cannot Exist, Without The Other, Dynamic Ergonomic Systems, Breed Community Or Anarchic Maelstrom - The Long Held Ancient Belief System, of Architecture, as the Majestic Monarch, of the Nine Muse Avocations, Has Partial Resonance, in Futurity of Technologically Advancing World - Cutting Edge Science, Dynamic Fusion Physics, Ecological Greening, Must Be Skillfully Introduced - Architecture as Macro-Physiology, A Gestating Womb, A Humane Place of Habitation, Must Imprint This Indelible Philosophy The Closer To Natural Models of Geometric Organization, The Stronger The Structure, The More Consistant, With The Environmental Laws of Nature - Internal Environmental Design, Potential Panacea, Replicated The Best Conditions, For Life To Prosper - Social Mortification, In All Forms, Equivalent To Civilization Scale, Mass Suicide - Abrogation of Such Social Stressors, The Plethora of Human Suffering, In Demographic Ghettoized Concentration, A Penitentiary Model, Imposed Without Due Process, Unwitting Inmates Incarcerated, On Prison Planet - The Populations Guilt or Innocence Irrelevant - Pushing Back The Darkness, Mute Silence Of Cave Dwelling Neanderthal - Pushed Into The Glacial Mountains, By Cro Magnon Hominids - Ostracized in Trail of Tears, To Wasteland of Subsistence, Awaiting Extinction, in Collective Fear and Self Loathing -
Defending Spiritual Warfare, With Cave Painting Superstition, Calling Upon Gods, That Will Not Save Them, The Amulet of Power, Cannot Protect Them, The Harshest Forces Of Nature, Leveled Against Them - Devolving Back, To Such Primitive Ways, The Resurrection of the Brutality, of Beast Within Us, Not a Fairy Tale, or a Childs Game, To Play With - The Promethean Fire Bestowed Upon Us, Made Useful, in a Thousand

Different Ways, Can Also Turn Infernal, And Seek To Immolate Us - Unobstructed Progress, Adaptation, Invention, Discovery, Are The Greatest Methods To Subdue, The Impulse To Regress Back Into The Cave - Competing With The Hibernating Bear, The Saber Tooth Tiger, The Odds Are Stacked Against Us, Without Protection, Of Advanced Civilization -To Organize, Teach, Clothe, Feed, Shelter and Guide Us - Spiritual Warfare, Met By Attaining Higher Spiritual Vibration, Greater Intellectual Awareness of Reality - Architecture in it's Construction, The Lessons Learned in the Building, Are The Fundamental Stepping Stones, of Industrialized Technological, Scientific Progress -

I Speak From Experience, Have Built Electric Railroads From Scratch, Introduced Patented Electronic Innovations, To Lessen Environmental Impact, Design Skyscrapers, The Vertical Form, Of Horizontal Railroad Technology, Have Created a Fusion Physics, To Innovate Their Engineering, and Bunker Stabilized Design - No Good Deed, Goes Unpunished, Obstructed and Assaulted, At Every Turn, Plagiarized of My Ideas, Discarded After Original Concepts, Were Highjacked and Bastardized - Bereft of Business, Design Credit, Renumeration of **ANY KIND**, Individually Targeted, by Corporatized Corruption - This Obviously, Is Not The Progress, I Am Referring Too - Having Hand Built, An Electric Railroad, With Overhead Catenary System - Technically Skilled, in All Forms, of Heavy Machine, Engineering Operation, Payloaders, Cranes, Backhoes, Bulldozers, Forklifts, Trucks, Trains, Using Taylorism Efficiency Modalities, Building a Functional Railroad, With a Selected Technical Crew, of SIX PEOPLE, = The Means Of Production Achieved, Through Mastering Mechanical Engineering - My Business Partner and I, Personally Training Them All, Design Building a Railroad, Upon the Streets of New York - Who Are The Neanderthal, Who The Cro-Magnon, What Method Of Progress, Empirical Innovation, or Industrial Espionage?, Will Be the Modus Operandi, That **HISTORICALLY BRANDS,** our Civilization: As Inventors or Pirates, Enslaved, Or Free Men? - Inquiring Philosophical Minds, Myself Included - Would Make Sense, of Such Things, That Approach The Diabolical, By Progressive, Oligarchic Degrees - Maul Hammer Rail Spiking, Into Solid Oak Railroad Ties, With Specialized Equipment Reused, Industrial Artifacts, From the 19th Century - I Dreamt of My Magnus Forefathers, Who Built This Country, Every Strike of the Hammer, an Inspiring Reminder

What Have We Here, To Show For It All - Fifteen Years Later, Destroying Physical Evidence, Of What Was Left Behind, Perpetrated By Corrupted City Agencies and Their Incompetent Bureaucratic Officials - A Plagiarized Skyscraper, A Mournful Blight To My Eyes, A Trans Hub Architecture, Left Behind, in Scornful Mimicry, Then Credited, To Another Architect, While Being, **IT, COINTELPRO**, "Character Assassinated" - I Speak With (Author)ity Upon Such Things - The Secret Alchemy, of Such Created, Engineering Feats, Phenomenal, Even In Facsimile, Are My Brain Children, My Inventions - The Reality of This, Will Be Revealed - I Would Swear And Testify To This, Bold Proof Displayed, Before You Now - The Literary Agency, By Which Repressed History, Is Most Often Displayed: Search, An Ye Shall Find, Further Evidence Compounding, Confirming Provenance, Accelerating Gravitas, of This Book's, Original Authenticity, and Fact Checked, Righteous Veracity - No More Time For Sportive Tricks, Shells Games, Bait And Switch - Watch The Accused Den of Thieves, Become Tongue-Tied Idiots, Pinned,

Punch Drunk, Against The Boxing Ring Ropes, of REALITY - Wagging Tongues, Signifying NOTHING - With Eyes Rolling, Rigged Bout Turning Sideways, Taking The Fall - My Business Partner, And Greatest Friend, Robert Diamond, Discoverer of the Atlantic Avenue Tunnel, (Also Robbed, By NYC, NatGEO, NYDOT) Chairmen and Founder, of BHRA, The Brooklyn Historic Railway Heritage Trolley, FedDOT Grant Project: www.brooklynrail.net - Fellow Researcher, Inventor Colleague, and Technical Project Manager, for the WTC Skyscraper Design, Coney Island Master Plan, Transportation Engineering Expert - Without Whom, This Book Would Have Never Been Written - I Owe Him An Immense Debt, of Gratitude Together We did Many Things, To Change The World - This Book Is Dedicated To Our Partnership and Friendship, This Is His Legacy, As Well as Mine - Let The Truth, Be Generally Known, Recorded Thus - I Tell This Story as a Cautionary Tale, Yet Encourage The Genius Inspired, To Follow Their Path - The Wisdomatic Do Not Stop And Wait, For Those To Follow, The Strike Ahead, Leaving A Beaten Path, To Their Invisible College - To Find The Greatest Inventions, Piled High To The Rafters - Inter-Related Systems, That Form Technological Matrix Synthetic Organism - Mimetic Empirical Observations, of Cornucopian Nature, Replicating Biomorphic Forms, With Synthetic Analogs - Choosing Architecture, as My Aesthetic Subject Matter, Railroads as First Industrial Revolution, After Gun Smithing - If You Can Build A Railroad, Requiring Every Known Technology, You Can Design Engineer and Construct, Anything - **"The Future is Yours, You Can Do Anything"**

Dominated, Manipulated and Abused, Morale Sunk Low, Our Projects Robbed, Or Else, Destroyed - I Took It Upon Myself, To **CONSTRUCT,** This Slim Volume, Form a Publisher Company, **NORDHAMMER,** To Achieve The Means of Production, Realizing, From Past Experience, in Intelligence Communities, The First Amendment, The Right To Free Speech, **IS STILL INTACT** - Perfect Archimedean Fulcrum, I Would Devise, To Loosen The Tongues, Of Those Who Have Plotted Against, Looted, and Destroyed My Business - Replaced By Another, **PUBLISHING WING -** That Cannot Be Tampered With, Quite So Easily - More Books Will Come, More Are On The Way - Mister Diamond and I, Are Quite Prolific Writers - Venting Our Spleen, An Acquired Taste, Which We Have Learned, to **RELISH,** as Fine Delicacy - Railroad "Superturnkey" Engineers, Railroaded, Out Of Our Own Business, Manipulated to Demonstrate, **THEN STEAL OUR INNOVATIVE IDEAS** - Railroading, Often Applied, to the Literary Arts - No Need To Resort, To Yellow Journalism - The Chronicle of Recorded Facts, Will Prove Sufficient, To Achieve "Locomotive Traction" - Transporting, Our Uncalled For Enemies, Hog Tied To Rusty Rails - I Suspect They Will Read This, With Arrogant Dismissal, Connivers, Scheming Away, At Their Last Machinations, On The Strategic Board - Surprise Attack Offense, Is Generally Met With Intensified Rage, of Amalgamated Defense - Undeclared Warfare, Is Terroristic in Nature, a Sign of **WEAKNESS**, The Recourse of Despots and Tyrants - In The End They Often Dangle From **PIANO** Wire, Being Spat Upon, as the War Disenfranchised, Pass in **CONTEMPT** - The Hegemony of The Big Lie, Brings The Mob To Frenzy - Demagogue and Propagandist Beware, **WHEN THE TRUTH COMES OUT -** No Bunker Deep Enough Will Hide You - The Truth Will Set Men Free - The Big Lie Will Bury You, By It's Sheer, **NEGATIVE GRAVITY**

Archimedean Lever Applied To Anything, With Sufficient Force, Will Send Obstructions, Of Any Size, In Opposite Direction - As Long As The Lever is **TRUTH,** and the Leveraging Force, Sufficient - Offset of Tangent, Moment of Sheer, Kinetic Motility, Axis Wedge Point, Homeostatic Tipping Lever, Avalanche Effect - Far More Effective, Than Bentham's Penitentiary Method - I Have Found Most Criminals, By Inherent Nature, Are Unrepentant, The Archimedean Lever of Ostracism, a Far Better Tool - Let Them Gather, In Glacier Mountain Caverns, Like Mute Neanderthal, Give Up Plagiarized Architecture, Take Up Primitive Cave Painting, Cave Dwelling, Simian Interior Design -

Let The Living Stone, Speak To Them, Like Echo, of Greek Legend, Let Them Gaze Into Cavern Pool, Like Narcissus, Blinded With Self Love, Inflated Egotism - Much Can Be Learned, in Meditative Hermitage, Prepare To Descend The Mountain, and Face Reality However I am Doubtful, After Nearly Thirteen Years of RICO ARC, Conspiratorial Silence, Anyone Among Them, Has Found "Religion", Or DISCOVERED "Honesty" - No One Has Reached Out, To Apologize, Ample Time, Contacts Made, The Only Response, The Awkward Silence, The Leering Gaze, The Tell Tale Hatred, The Arrogant Smirk, Of Corrupting Influence, The Unjust Rise To Power, of the Professional Criminal In Point, of Fact, Am Somewhat Gladdened, That Such a Tragic, Chain of Events, Has Led Me, Irresistibly, To Form The **NORDHAMMER PUBLISHING COMPANY** - A Linguistic Repository **ARCHIVE**, Moved Slowly Into, **FREE PRESS**, Philosophical Dialectical Forum, New Lyceum, Where We Can Share, our **COPYRIGHTED** Technology, Genius Innovations, and Original Inventions, With the, Rest Of the World, In Unobstructed **PEACE** - (**BEYOND, Benthamist Penitentiary, of NYC)** - The Greatest Way I Know, To Beat Back: **THESE POISONOUS SNAKES AND SLITHERING, "FORKED TONGUE" REPTILES**, Out of The Public Park Grass, Where Innocent Children Play - Perhaps Painting, an Unfortunately, **ACCURATE** Portrait, of New York City, in 2015 - Place and Time, Suffering From Grandiose Delusions, of Corporatized, Self Entitlement - Peculiar, to the Nuveau Riche Pretentiousness, and Contemptuous Arrogance, of Rising Oligarchies - Self Defined, Ignoble Aristocrats, Robber Barons, As Best Described - Oblivious To Everything, Beyond Themselves, Egotistical Gluttony, With Piratical Consuming, Conspicuous Obese Appetite, That Knows No End - Creating a Delicacy of Righteousness, **That THEY Will Choke Upon Feeding, Bon Apetite - Eat it All! = "Like Casting Pearls Before SWINE"**

Indomitable in Careering Spirit, I Will Not Halt or Yield - Righteous Truth, My Armor, My Philosophical Thoughts, My Sword - Come From Unbroken Line, of Scot Boernician Clan Warriors, Turner's, Lamont's, MacDonald's, Guardians of the Throne of Scone, From, Ulster Lawmen Kings, My Ancestral Will, and Proverbial Calling - Judge Not A Book, By It's Cover, Nor The Semblance of a Man, Without Fathoming, His Thoughts - Never Underestimate, Those You Have Wronged, Nor Compound The Trespass, With Chronicle of Abuse - Eventually, You Will Meet Your Match, Jousting With Unbated Lance - From Armoric Knights, My Ancestors Rose, We Fought The Romans Back, To Hadrian's Wall - Built To Keep Us Out, From Angle-Danish Land, To Contain The Ferocity, of Mountain Descending Highlanders - Giant Claymores Strapped, Across Our Backs, To Penetrate, The Roman Legions, Cutting them Down, From Chops to Reigns,

Our Soulmate Woman Followed, in Legion Killing Train - No Prisons Taken, No Quarter Given, Invasion of the Highlands, Remains Unforgivable - Taking No Hostages, No Need For Roman Slaves, Dwarves of the South, Anathema To Ancient Viking, Celtic, Pict, Barbarian/Boernician Ways - So Too, Basque Lineage, Undefeated Mountain People, Fearlessly Independent, Sent Gaulish Frank Carolingians Packing, The Song of Roland, His Trumpet Sounding, Sole Survivor, of Basque Land Invasion - Now You Have Meddled in My Affairs, Rebuked My Honor, Stole What I Possessed - Thinking Me Long Gone, You Built My Designs, Proving Beyond Doubt, You Are Thieving Plagiarists I Think, I Hear The Song of Roland, Trumpeting Retreat, From the BASQUE Pyrenees - After Saracen Lowland Massacre, Emboldened by Triumph, Facing Horn Blowing Defeat - Though Clarion Siren Warning, Filled the French Valley Air Below, No One Turned Round, To Face The Music, To Ever Challenge The BASQUE AGAIN - No Wonder The Epic Poem, Remains Anonymous (Embarrassing!) Your Hadrian's Wall, Will Not Contain My Wrath - (My People, Do Not Go Gently, Into That Good Night - We Prefer To Send Other To It) It Is I, Who Designed It, You Who Stole It - Who Is The Engineer, Equipped With Siege Engine Technology - Who Is The Peddler, Tinker, Wandering Scab, Dealing In Stolen Property, He Has No Provenance For - 'An Empty Suit, and an Architects Stamp, Upon Stolen Plans, Does Not An Architect Make', My Kleptomaniac Friends - Nor Can They Magically Turn You, Into Professional Writers, To Defend Themselves, With The Poison Pen - I Do Not Count Polymaths, Geniuses, or Intellectuals, Among Their Pirates Crew - The NEED to Steal from me, would have Become irrelevant, if they could design their Own skyscraper, would not need four years Of outboard architectural engineering, to Alter and obfuscate the original design, SOM Would be **SINGING THEIR OWN PRAISES** - At Completing the World Trade Center, And MTA Transit Hub, Instead of Remaining **DEATHLY SILENT** - Like The (**Schrodinger's**) Cat Who Ate, The Proverbial Canary: Yellow Feathers Pluming, From Their Lying Mouths - Perhaps They Should Hire A Stand In, "Spin Doctor" Ventriloquist, To Answer The Questions, of the, **International Press** - Better Than Having, The Canary Feathers Flying, When Asked Pointed Questions, About Their Plagiaristic, Charlatan Roles, Blaming Each Other, In The "Condemning Process", All The While Canary Feathers Spewing Forth, To Camouflage the Media Circus, They May Consider: **CHICKEN LINE DANCING, IN CANARY SUITS, Just Act The Impostor Imbecile**

A Web of Lies, A Multi-Billion Dollar, Architectural, "Shell Game" Cut and Paste, Swaying Fabrication, of a Twin Building Skyscraper, Hackneyed Into One Monstrosity Four Years of Billable Architectural Office Time, Forty Man Staff, Working Full Time, Creating the Great Metamorphosis, **BARING STRIKING RESEMBLANCE**, Like Siamese Conjoined Twins, Of My Original Skyscraper Design - A Pathetic Rendition, I Retro-Engineered, in Five Minutes, Showing the Forensic Trajectory, of the Plagiaristic Dissimulation, Being Passed Off, as Original Architecture - And What of the Internal Engineering, Stealth Hidden Throughout, The Entire Construction Process - What Could They Be Hiding, Behind Stage Magician Curtains? - Time Lapse Video, of the Shard Building Construction, Shows My Architectural Engineering, Skyscraper System - From Top to Bottom, A Plagiarism, of A Plagiarism, Redundant Design Work, From The WTC, Building One - CSUK, Found Their Architect Stooge, in Piano - The "Gigantic Elevator Bank", Served as "Makeshift, Sway Arrester" - Stripping The

Buttressed, Isotonically Tensioned Components, of the Original Design, They Achieved Brilliant Architectural Engineering Solution, Thirty to Fifty Percent, Interior Space Allocation, Dedicated To Elevated Shafts - Well Done, TWICE IN A ROW = Flawless! -

Perhaps They Should Transition, There Considerable Talents, to "Bunker Like" Outhouse Construction Modalities, Cement Reinforced Dog Houses, of Benthamite Concrete Cast, Penitentiaries - Perhaps Their Construction Techniques Will Catch On, Building Luxury Highrise, Penal Colonies (Like in NYC) "The Residential" Section, Intended for the upper floors, of the original WTC design, segregated out, Transplant Built, in Midtown Manhattan, by CSUK - **BRAVO** - Another Soaring Triumph, of Architectural Banality - They Are On Quite a Design Roll, **Downward Spiral** - My Guess, They Have No More Original Designs, To Steal? Pick The Bones And Feathers, OFF THE CANARY, Three Times in a Row, That is **NOT CRICKET**, CSUK - LOL

Upon More Positive Ending Note, Vernacular Architecture as Social Modality of Community Building - The Pennsylvania Dutch, The Amish, Have A Tradition of Barn Raising, for Members of the Community - A Barn Structure, is Labor Intensive, Requires Great Quantities, of Building Material, And Must Be Made, To Industrial Grade Specification - The Amish, Experts in Carpentry, Have Streamlined the Construction Process, to Several Days - Using Traditional Methods, Electronic Machinery Prohibited, Joist Beams, Drilled and Doweled Connections, Built To Withstand Centuries, of Continuous Agricultural Use - The Community Building Process, Builds Community Social Bonding, One of the Most Enduring Religious Communities, in North America - Ergonomic Dynamics, Self Sufficient Land Allotments, Gathering Places, For Organized Social Activity - Essential Components, for Viable and Humane, Community Habitation, Demographically Calculated, to Utopian Rural Scale - Another Example of Organic Human Scale of Proxemics, Autonomous Zone of Privacy, That Can Be Adopted Geometrically, To Form Agricultural, Philosophical, Scientific Rural Communities, Like Findhorn, Scotland, Soleri's, Arizona Community - Forerunners, of a New Architectural Constellation, Best Suited, To Urbane-Humane, Ergonomically Dynamic, Community Based, Utopian Modality, Habitation - The Diaspora Jewish Communities of the Russian Steppes, Situated in Czarist Controlled, Cossack Territory, Experienced Constant Harassment, Unwilling To Conscript, into the Russian Military, and were Often Raided By The Cossack Mercenaries, of the Czar - These Villages, Developed Over Time, Had Excellent Ergonomic Dynamic Scale, Stable Population Growth, Skilled Work Force, Independent Self Sufficiency - To Ward Off The Cossack Attacks, Which They Also Considered Partially, a Metaphysical, Spiritual Warfare Issue - They Employed Arranged Marriage Matchmaking, To Create Astrological Synastry, Raise Psychopomp Highest Vibration, Of Spiritual Amplitude, Drawing Down, **The SHEKINAH** of Indwelling, The Comforter, Who Would Grace and Bless Righteous Love Matches, Metaphysically Protecting The Community - This Traditional Process, of Matchmaking, Produced the Desire Effect - Like The Gollum, of Krakow, Rabbinically Created Sentinel, To Guard The Community - The Matchmaking, Produced a "Protective Barrier", an Autonomous Creative Zone, Which Helped the Communities, to Prosper and Endure, Even During The Czarist Cossack Mercenary Period - Indicating Proper Astrology Arranged, Synastry Marriages, Increased a Synergistic Human Energy Field, Rebuking Evil

Produced Positive Synchronicity Outcomes, Enhancing The Utopian Paradigm, of People Living, in Optimal Coefficient Righteous Community - With Great Success, Even During Stressor Times, of Demographic Pressure, The Community Could Self Regulate, Autonomously, Within Hostile Territory - Indicating The Efficacy of Human Synergistic Energy, Through Soulmate Matchmaking, Another Critical Factor, in Idealized Gnostic Community Constellation, Architectural Modalities, Should Be Cognizant of, as an Additional Human, Synergistic Power Factor - The Design Modulus, of Ergonomic Scale, Between Societal Interaction, and Individual, or Marital Privacy, Schematized to Produce, The Greatest Fusion Physics Quantum, of Collectivized Community Energy -

Before The Shamanistic Vision, and Subsequent Creation of the Peace Pipe, The Native American Population, Was Formed in Loose Migratory, Subsistent Populations - When The Prophetic Visions, of Wild Tobacco Cultivation, Domestication Processing And Age Curing, The Building of Sacred Pipes, With Winnebago Minnesota Pipestone, The Ancient Sacred Prayers, That Bless The Tobacco, Into Traditional Indian Medicine So Powerful an Influence, That These Tools of Ascent Had - That Two Major Confederations Formed, The Algonquin of the North, and the Cherokee Confederacy, Of The South, Consisting of MILLIONS of Individuals, Congregating Bands, That Formed Democratic Civilization - The Algonquin Long House, The Passing of the "Talking Stick" of Free Speech in Political Counsel, Great Inspiration, to the Revolutionary Founding Fathers, Who Were Fascinated, By Their Aboriginal Form of Democratic Government - The Boston Tea Party, an "Honorific", To Their Mohawk Tutors, in Political Science - The Oldest Continuous Democracy Upon The Planet, Centered Around The Sacramental Use of the Sacred Peace Pipe, and Blessed Medicine Tobacco - The Norman Parliament, of Sicily, the other most enduring Democracy, However Relatively Speaking, a House of Lords, Rather Than a Tribal Community, Direct Democracy, Closer to the Athenian Polis Model - The Algonquin Mohawk Longhouse, Built To Accommodate, an Entire Political Party, Like Their Viking Analogue, A Place of Group Decision Making, A Centralized Hall, Meeting Place, A Royal Court Where Laws, Were Judged and Made - Both Community Hubs, For Fiercely Independent, Warrior Peoples - Both Successful, In Organizing Them, Into Multi-Million, Member Empires, Controlling Vast Regions, of Geo-Political Influence - Like The Viking Long Ship, Light Enough, To Carry Like A War Canoe, Across Miles of Land, Strong Enough, To Cross Oceanic Floods, With Armored Warriors, Heavy War Horse Cavalry - Ergonomic Scale, Ingenuity, and Social Community Building, Mobility - Allowed These **HIGHLY** Independent Groups, To Unite, Into Community, Syndicating Their Group Modality, Across Vast Empire Building Demographics - Thus Creating: "Barbarian-Civilization" - Architectural **INCLUSIONISM**, Designed Into Their Centralized Gathering Point, Vernacular Structures, Common Denominator, Community Building Thread - Habitational Architecture, Promoting Civilizing Function, Social Organization, Free Dialectical Discourse, Through Parliamentary Procedure, To Engineer Group Cohesion and Expansionism, From Otherwise Independent, Tribal Warring Factions - Commonality of General Purpose, Seems The Other Cohesion Factor, From Metaphysical and Religious Communities, To Barbarian Civilizations, and Everything In Between - Shared Vision, Matched Purpose, Driving Force, of Community Building:

As I Lay Dying, In The Years To Come, Perhaps Philosophically Reflecting, Upon What Has Come and Gone - I Will Take What Precious Time I Have, To Reflect Upon The Ancestors, Who Have Come Before Me, The Warriors, The Soldiers, The Artists, Metaphysicians, The Philosophers, The Teachers, The Doctors, The Architects, The Designers, The Engineers, The Intelligence Men, Genealogical Roots, That Spread Millennium - Their Genius Shining Through, The Hundreds of Generations - The Community, That Formed Around Me, That I Now Speak Upon, With The Certitude, of Eloquence - The Knowledge, I Inherited From Them, Which is the Greatest Treasure, Family Jewel, Of Family Legacy - Will Think Kindly Upon Those, Who Became My Second Family, Surrogate Parents, Professors, Teachers, Loyal, Life Long Friends - Who Graced My Own Life, And Further Strengthened, My Indomitable Spirit, To Overcome, Humblest Beginnings - Who Encouraged Me To Speak My Mind, Express My Creativity, Taught Me, The Value of Knowledge, Fair Play, and Ethical Truth Seeking Without Them, The Composition of This Book, Would Be Impossible, to Elucidate, From Subject To Context, To The Original Architecture, Physics and Philosophy, Herein Contained - It Echoes Their Collective Wisdom, and Thorough Righteousness - They Saw, The Spark of Intelligence in Me, Imparted A Fantastic Knowledge Odyssey, With Flawless Pedagogical Altruism, Learned Unwavering Truth, Empirical Apperception, Psychic Gifts - Many Have Passed On, Some in Old Age, Others Cut Down, in Midlife Tragically, Feel Them Near Now, When Speaking Upon These Things - Feel They Would Relate To This, Like a Message, in a Bottle, Upon What This Volume Contains - The Fluidity, of Their Inherited Knowledge, Now Passed On, To The Next Generation - Perhaps In Part, a Family Bible, That Bares No Names - Would Rather, They Remain Anonymous, Those Who Still Live, I Would Spare, From This Unwholesome Tragedy: My Own Cross To Bare - For Those Coming Up, I Would Warn Them, Of Such Treachery - Let Them Know The Signs, Teach Them To Protect Themselves, From Intellectual Feudalism, and Plagiaristic Blight - Sometimes Expressed Best, In The Confines of A Book - Best Method Of All, To Express, The Most Complex, Compendious Ideas, Carefully Articulated, into Sensible Form - So That Aspects Can Be Studied, Synthesized, Or Cast Off - Therefore, For The Living And The Dead, For MY People, That, I Write This For - Too Soon From Dust To Dust, We Are Carried Forth - Part Eulogy, Part Epitaph, Part Living Manifesto, Existing Now, In Eternity - Having Moved Slowly Into Print, The Masterful Advise, of Borges, Argentinian Homer, Kindred Spirit - Weighed My Words, Exacted My Arguments, Correcting, The Sophist Cacophony, Heard All Around Me - Clarification of Reality, Adjustment of Propagandist False History, Has Inspired Me, To Take These Wicked Men, To Account - I Will Not Live A Lie, Even Those Created By Another - I Would Thus Testify, In Open Forum, Let The Chips of Corruption, Fall Where They May - My Original Skyscraper, Had a Great Plagiaristic Fall, from Inspired Grace, Yet Another Compounded Lie, Repeated Insult, To Chronic Injury - Like Humpty Dumpty, Who Fell Off His Wall, All The Oligarchs Architects, All The, "Would Be Kings", Structural Engineers, Couldn't Put My World Trade Center Design, Back Together Again - Still Waving In The Wind, Like an **Eggman**, Upon A Wall, Gargantuan Baffoonery, Wobbling, With Motion Sickness: Monstrous Frankenstein Effigy, of it's Collaborative "Creators", In a Butchering, **"Chop Shop"** I Accept No Blame, For Derivative, Pseudo-Engineered Design, "Back Street Abortions" Just The Stolen **"Exquisite Corpse"** Though Not "Well Hung" = Grave Robber Derived, From

"Swaying" Gallows Pole, From My Original, 2002, Twin Obelisk DIG/BHRA, Skyscraper Design Plan - Like Maggots Crawling, Over Wasted Side, of Fetid Beef - Like Arch Villains, in a Shakespearean Play, Their Conscience, Ghost Haunts Them, With Monologue Asides, Fathoming, The Innermost Depths, of Their Depravity, During Plotting Prevarication, Or When, Their Horrific Catalog of Crime, Is Theatrically "Revealed" - I Would See, Such A Modern Variation, on a Theme, Transpire - With the Cotton Mouthed Choppiness, and Cryptic Allusion, To Alleged Crimes, They Never Dreamed, Would Lamb Baste Them, in the Press - Skewered and Rotating On a Spit, Over White Hot Coals, Worm Ridden Apple, Wedged in Their Gaping Mouths Like The "Mugwump" Barbeque, in Burrough's, "Naked Lunch": Pass The BBQ Sauce -

People, That I Have Mentioned, cc'ed, on My "Whom It May Concern" Letter, And Yes, It Concerns Me Greatly - By Nature are Reclusive Animals, Preferring the Solitude of Ivory Tower, The Boardroom Conspiracy of Power Brokers, Secrecy and Silence, The Tools of Their Nefarious, Stock and Trade - An Honest Word, Never Passing Their Lips, Their Conscience Pronounced Dead, Years Ago - Reflection Upon Their Crimes, A Source of False Bravado, Boys Club Sardonic Humor, Like Headhunters Bragging Rights, Over Missionary Decapitations - During Heart Attack, Steak Dinners, Scotch and Martini SWILLING, They Become Ham Thespians, and Impersonate Their Latest Victims - Patting Each Other On The Back, Like Rhodes Scholars, or Nobel Laureates - They Reinforce Each Others Plagiarisms, Some Admiring, The Violation of Good Will, Others, The Con Artistry of Solicitation, Some the Fidelity, to the Letter of the Law, By Which They, Vampirically Function - Other Praise The Work Itself, How Much Money It Will Fetch, On The Open Market - How Much Kickback They Will All Receive - Calculated Down, To The, **LAST PENNY** - Some Consider the Logistics, How Many Glaring Facts, Must Be Messaged, How Many "Stand In" Impostors, Must Be Positioned, To Take The Heat, If "Things Go Sideways" - The Media-Centric, Begin The General Call, Hailstorm, To "Spin Control" The Press, Toyed With, Like **The Russian PRAVDA -** And If That All Fails, Their is Always, the "Fallback Position", To Corrupt City Politicians, and Their Guard Dog, "Payola" Police Force - Envelopes are Stuffed Full, Like Christmas Stockings, With Care - All Contingencies Considered - Over Single Malt Scotch, and Five Pounder, Prime Rib Dinners - Instead of Goodbye Handshakes, They Give Each Other, Courtesy "Reach Arounds" - Would Rather Eat With Cannibals, They Have Better Table Manners - Socialize With Untouchables, In Leper Colony - Then Waste My Spit, Aimed, In Their General Direction - I Would Rather Describe, Their Oligarch Antics, As It Amuses Me - At Least I Got That All, Out Of My System, Think I Need To End This Now, Smoke My Shamanic Peace Pipe, With Blessed Medicine, Spiritually Meditate - Pray To The Creator, Never To Become, Like These Excuses, For Human Beings - Bad Enough, I Had To Deal, With A Few, Too Many: Formaldehyde Specimens, Midway Sideshow, Freak Zoology, Pickled In Their Own, Poisonous Juices - This A Field Guide, To Identify, Their Corrupted, Human Species - Build Immunity, Through Righteous Prayer, Tell The Absolute Truth, Don't Ever Sell Out Your Artistic Vision, Formulate Your Own Philosophy Remain An Individual No Matter What, Sell Your Soul To No One: Hell Is Filled With Liars, Stay Close To The Absolute Center: "The Future Is Yours, You Can Do Anything"

Beware, Have Heightened Awareness, Regarding, Coercive Intelligence Programming, (COINTELPRO) Mind Kontrol Ultra (MKULTRA) Nazi Paperclip Psychiatric Artifacts, Propagandist, Anti-Philosophy, Endemic To The Architecture, of Social Mortification, Formerly Conceived, as Benthamist Penitentiary, Penology Systemics: The Cult Hive Mind, Antithesis, of Righteous Community Configuration - Mixture of Psychotropic Medications, Hallucinogens, Deliriants, Debilitating Agents, Used in Weaponized Form To Pacify and Control, Large Urban Populations, Enemy Troops, in Theaters of War - Civil Unrest, Riot Control: BZ, Agent 15, The Worst Offenders, of Them All - Fluoridation Of Water Supplies, Targeted the Calcification, of the Pineal Gland, "The Third Eye" Brain Organelle, of Psychic Apperception - The Other Brain and Central Nervous System Neuroleptics, Targeting the Limbic: "Fight of Flight System" - Controlled by the Master Twin Brain Gland, of the Hippocampus, Producing Synaptic Non Conductivity, Cortisol And Adrenochrome Arousal, General Anxiety Disorder, PTSD, Extended Exposure Produce, Morbidity Through Smooth Muscle, Cardiopulmonary, Respiratory Renal Failure - Acetylcholine Levels, L-Tyrosine, Essential in CNS Synaptic Function Are Compromised, Through Hippocampus Atrophied Function - Choline, Pharma Gaba, Garcinia Cambogia, are Advisable Prophylactics, Against Such Neurological Psychotropic, MKULTRA Agents - Melatonin, a Remedy For Pineal Gland Calcification - Natural Supplements, To Restore Optimal Mental Health and Well Being - Fluoridated Water, First Used In Austwitz Concentration Camp, Like Ziklon B, Wafferin Blood Thinners, in the Notorious Concentration Camp Shower Heads - Such Coersive Use of Chemical Warfare, Produced Willing Participants, and Pacified Victims, of Nazi Holocaust - This Is Of Course, The Extreme Case, of the Penitentiary Work Slave Model, The Hive Mind Brain Child of Corporitized War Mongering Fascist State - The Antithesis and Antipode, To Righteous Equalitarian Community, Direct Democracy, Scientific and Philosophical Meritocracy - Pharmaceutical Industrial Complex, vs, Holistic Agricultural Self Sufficiency, Yet Another Glaring Example, of Penitentiary vs, Community Habitation - Pyramidal Hieratic Oligarchy, vs, Flattened Hieratic of Equalitarian Polis Community The Architectural Modality Reflect These Things, Slave Factory Sweathouse, vs, Ergonomic Demographic Synergy - One An Industrial Slaughterhouse, A Human Stockyard, Extruding Murder and Mayhem - Human Medical Experimentation, To Increase Holocaustic Efficiency - The Other A Utopian Panacea, Set To Human Ergonomic Scale, Where Stabilized Population, in Autonomous and Congregate Interaction, is Echoed in Architectural Scale, Proportion, and Organic Proxemics Relativity - One Controlled by Medically Unethical, Militarized Psychiatry, The Other Led By Wisdomatic, Independent Self Determination, Guided by Righteous Philosophers, Shamans and Healers, Optimizing Longevity, Procreation, Community Social Integrity - One Gravitates The Wicked, Corrupts The Otherwise Good, To Abominable Inhuman Acts, The Other, Constellates Soulmates, Engenders Good, Proliferates, Invention and Creativity - The Herd Impulse, In Human Beings Is Stronger Than Men Thing, The Gestalt of Place and Time, Reduce Mankind To Locust Maelstrom Swarming - Architecture and Urban Planning, is the Mechanism, For Either Outcome - Therefore Plagiarism, of it's Original Design and Intent, Begets Cataclysmic Outcomes, Dire Repercussion: Beware of Ivory Towers, Sacrificial Temples, Babylonian Spires They Are The Guard Towers, of the Penitentiary, All Are Prisoners, All Are Warders -

## LE PREMIER LIVRE.

HRE- CV- LES GAL- LICVS        LE HER- CV- LES FRAN- COIS.

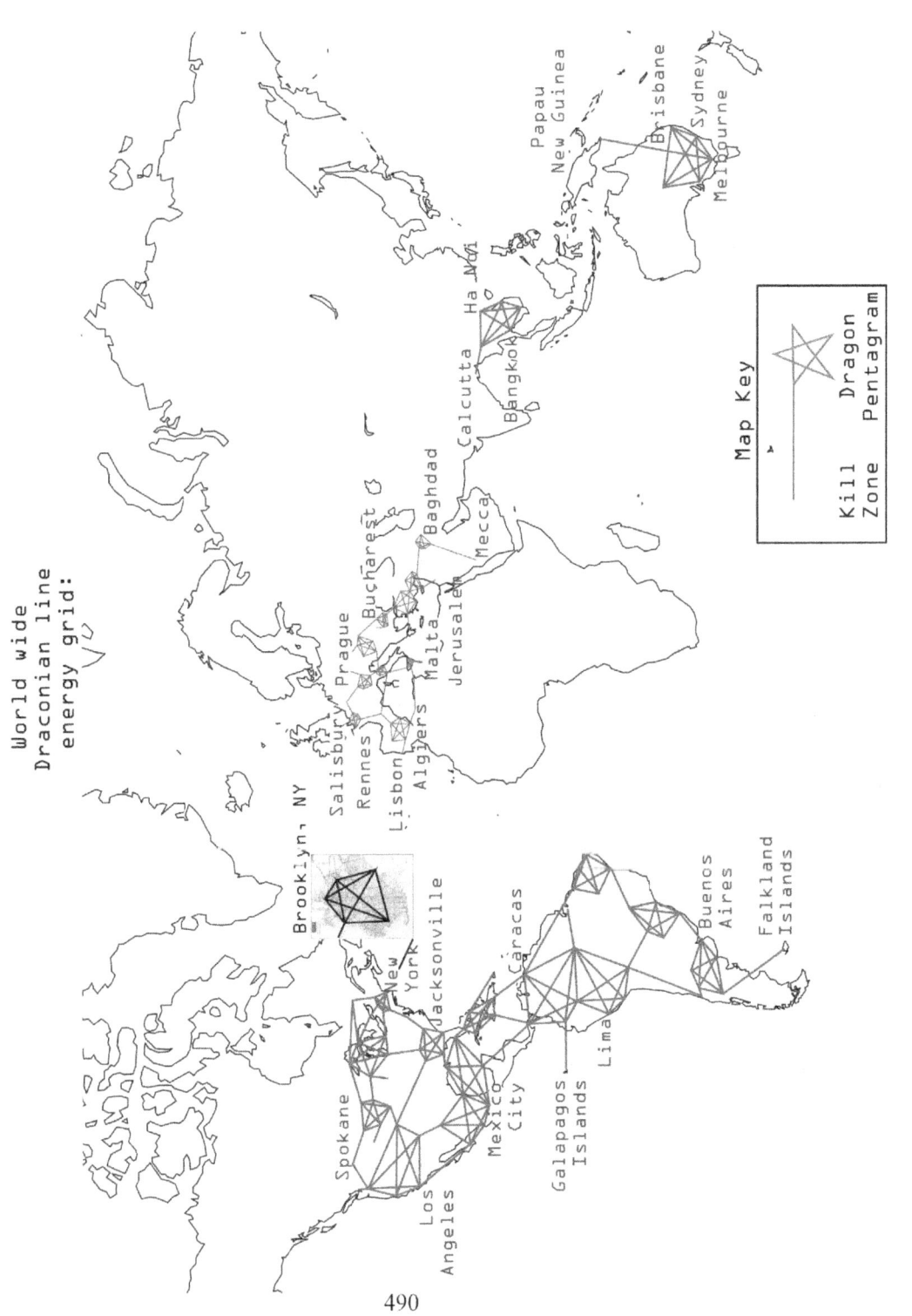

**NORDHAMMER PUBLISHERS**

www.ingramcontent.com/pod-product-compliance
Lightning Source LLC
Chambersburg PA
CBHW080934170526
45158CB00008B/2289